QUANTUM STATISTICAL MECHANICS

DOCUMENTS ON MODERN PHYSICS

Edited by

ELLIOTT W. MONTROLL, Institute of Defense Analyses
GEORGE H. VINEYARD, Brookhaven National Laboratory
MAURICE LÉVY, Université de Paris

A. ABRAGAM,
L'Effet Mössbauer
H. BACRY,
Lectures on Group Theory
K. G. BUDDEN,
Lectures on Magnetoionic Theory
J. W. CHAMBERLAIN,
Motion of Charged Particles in the Earth's Magnetic Field
S. CHAPMAN,
Solar Plasma and Geomagnetism and Aurora
H. CHIU
Neutrino Astrophysics
A. H. COTTRELL
Theory of Crystal Dislocations
R. H. DALITZ
The Quark Model for Elementary Particles
BRYCE S. DeWITT
Dynamical Theory of Groups and Fields
R. H. DICKE
The Theoretical Significance of Experimental Relativity
M. GOURDIN
Laws of Symmetry Theorem of T. C. P.
D. HESTENES
Space-Time Algebras
JOHN G. KIRKWOOD
Selected Topics in Statistical Mechanics
JOHN G. KIRKWOOD
Macromolecules
JOHN G. KIRKWOOD
Theory of Liquids
JOHN G. KIRKWOOD
Theory of Solutions
JOHN G. KIRKWOOD
Proteins
JOHN G. KIRKWOOD
Quantum Statistics and Cooperative Phenomena
JOHN G. KIRKWOOD
Shock Waves
JOHN G. KIRKWOOD
Dielectrics--Intermolecular Forces--Optical Rotation
V. KOURGANOFF
Stellar Nuclear Reactions and Internal Structure
Paul H. E. MEIJER
Quantum Statistical Mechanics
A. B. PIPPARD
The Dynamics of Conduction Electrons
L. SCHWARTZ
Applications of Distributions to the Theory of Elementary Particles in Quantum Mechanics
M. TINKHAM
Superconductivity
R. LATTÈS
Methods of Resolutions of Some Boundary Problems in Mathematical Physics

Additional volumes in preparation

QUANTUM STATISTICAL MECHANICS

Paul H. E. Meijer

The Catholic University
Washington, D. C.

GORDON AND BREACH Science Publishers
NEW YORK LONDON PARIS

150 Fifth Avenue, New York, N.Y. 10011

for Great Britian and Europe:

Gordon and Breach, Science Publishers, Ltd.
61 Carey Street, Chancery Lane
London WC2, England

Library of Congress Catalog Card Number: 66-30555

Distribution en France par:

Dunod Editeur
92 Rue Bonaparte
Paris 6, France

Distributed in Canada by:

The Ryerson Press
299 Queen Street West
Toronto 2B, Ontario

Printed in the United States of America

Documents on Modern Physics

Seventy years ago when the fraternity of physicists was smaller than the audience at a weekly physics colloquium in a major university a J. Willard Gibbs could, after ten years of thought, summarize his ideas on a subject in a few monumental papers or in a classic treatise. His competition did not intimidate him into a muddled correspondence with his favorite editor nor did it occur to his colleagues that their own progress was retarded by his leisurely publication schedule.

Today the dramatic phase of a new branch of physics spans less than a decade and subsides before the definitive treatise is published. Moreover modern physics is an extremely interconnected discipline and the busy practitioner of one of its branches must be kept aware of break-throughs in other areas. An expository literature which is clear and timely is needed to relieve him of the burden of wading through tentative and hastily written papers scattered in many journals.

To this end we have undertaken the editing of a new series, entitled Documents on Modern Physics, which will make available selected reviews, lecture notes, conference proceedings, and important collections of papers in branches of physics of special current interest. Complete coverage of a field will not be a primary aim. Rather, we will emphasize readability, speed of publication and importance to students and research workers. The books will appear in low-cost paper-covered editions, as well as in cloth covers. The scope will be broad, the style informal.

From time to time, older branches of physics come alive again, and forgotten writings acquire relevance to recent developments. We expect to make a number of such works available by including them in this series along with new works.

ELLIOT MONTROLL
GEORGE H. VINEYARD
MAURICE LÉVY

INTRODUCTION

The following notes are the result of summer school conducted for the Greater Washington area at the Physics Department of the Catholic University of America. The course was intended for students which had adequate knowledge of quantum mechanics and statistical mechanics but not of quantum statistical mechanics as well as for researchers who were curious about the workings of the diagrammatical and related methods in quantum statistical mechanics. The course consists of four parts.

(1) A general introduction about the density function respectively density matrix with a number explicite examples. The Wigner distribution function, an exposé of second quantization and finally the diagrammatic method in the form as presented by Bloch and de Dominicis.

(2) The Green's function method is extensively reviewed and examples are given for an ideal gas and a gas with vander Waals-like properties; the result of a combination of the Green's function method and the Hartree approximation.

The second part of this chapter is devoted to the theory of the electron gas in particular the Montroll Ward Theory.

(3) The third chapter is almost exclusively describing the potentialities of the diagrammatical method. It describes the different types of diagrams that are in use, illustrates how the proof of the virial expansion expression can be obtained in an almost entirely pictorial way, in the classical case and then compares how the classical diagrams are generalized to the quantum mechanical diagrams and how one can go from the "vertical diagrams" to the torons of Montroll and Ward and to the "loop-like" diagrams of Bloch et al.

(4) The last set of lectures deals with the Pauli or master equation. Using the linear tetradic operator L an exact expression is derived for the time derivative of the diagonal elements of the density matrix.

The most important subjects missing are the Bose Einstein condensation and the superconductivity obviously impossible to cover in the time available.

We would like to thank our guest lecturer: Dr. R. W. Zwanzig of the National Bureau of Standards for his participation and acknowledge in general the support to our program by the Air Force; Grant AF-49(638)-452.

CONTENTS

Density Matrix

Paul H. E. Meijer

1 DENSITY FUNCTION

$$\rho(xx') = \psi(x)\ \bar{\psi}(x') \tag{1-1}$$

or in Dirac notation $|x\rangle\langle x'|$ hence the second label refers to the complex conjugate wave function.

The diagonal elements are the probability of finding a particle as used in the Born interpretation of quantum mechanics.

If we take the Fourier transformation we obtain a matrix

$$\rho_{kk'} = \int \rho(xx')\ e^{ikx}\, e^{-ik'x'} dx\, dx' \tag{1-2}$$

The original function could also be interpreted as a matrix with continuous indices.

2 LIOUVILLE EQUATION OR EQUATION OF MOTION

$$H_x \psi(x) = i\hbar \frac{\partial \psi(x)}{\partial t} \quad \text{times } \bar{\psi}(x') \text{ from the right}$$

$$H_{x'} \bar{\psi}(x') = -\, i\hbar \frac{\partial \bar{\psi}(x')}{\partial t} \quad \text{times } \psi(x) \text{ from the left}$$

$$(-$$

$$H_x \psi(x) \bar{\psi}(x') - \psi(x)\ H_{x'} \bar{\psi}(x') = i\hbar \frac{\partial}{\partial t}\ \psi(x) \bar{\psi}(x')$$

If H_x is replaced by $H(xx'') = H(x'')\,\delta(x-x'')$ then the first term is a matrix product.

$$\int H(xx'')\,\rho(x''x')\,dx'' = \int H(x'')\,\delta(x - x'')\,\rho(x''\,x')\,dx''$$
$$= H(x)\,\rho(xx') \qquad (2\text{-}1)$$

The second term is

$$\begin{aligned}\psi(x)\,H(x')\,\bar{\psi}(x') &= \int \psi(x)\,H(x'')\,\bar{\psi}(x'')\,\delta(x''-x')\,\delta x'' \\ &= \int \psi(x)\,\bar{\psi}(x'')\,H(x'')\,\delta(x''-x')\,\delta x'' \\ &= \int \psi(x)\,\bar{\psi}(x'')\,H(x''\,x')\,dx'' \\ &= \rho\,H. \qquad (2\text{-}2)\end{aligned}$$

In Dirac notation:

$$H|x\rangle = i\hbar\frac{\partial}{\partial t}|x\rangle \quad \ldots \text{ times } \langle x'| \text{ on the right}$$
$$\langle x'|H = -i\hbar\frac{\partial}{\partial t}\langle x'| \quad \ldots \text{ times } |x\rangle$$
$$\underline{\qquad\qquad\qquad\qquad}(\text{— on the left.}$$
$$H|x\rangle\langle x'| - |x\rangle\langle x'|\,H = i\hbar\frac{\partial}{\partial t}\,|x\rangle\langle x'|$$
$$\text{or } [H, \rho] = i\hbar\frac{\partial\rho}{\partial t} \qquad (2\text{-}3)$$

3 PURE STATES VERSUS MIXED STATES

$$\psi = \sum_n c_n\,\varphi_n \qquad (3\text{-}1)$$

describes the "state" of one system. Take many copies, labeled by $\lambda = 1, 2, \ldots N$. We have a set of functions

$$\psi^{(\lambda)} = \sum_n c_n^{(\lambda)}\,\varphi_n\,. \qquad (3\text{-}2)$$

Construct the corresponding density matrices

$$\begin{aligned}\rho^{(\lambda)}(xx') &= \sum_{n,m} c_n^{\lambda}\,\bar{c}_m^{\lambda}\;\varphi_n(x)\,\bar{\varphi}_m(x') \\ &= \sum_{n,m} c_n^{\lambda}\,\bar{c}_m^{\lambda}\;\rho_{nm}(xx') \qquad (3\text{-}3)\end{aligned}$$

We introduce the average over all systems which we call the density matrix for a mixed state.

$$\frac{1}{N}\sum_{\lambda=1}^{N}\rho^{\lambda}(xx') = \rho(xx') \tag{3-4}$$

The coefficients of this matrix with respect to a certain orthonormal function system [in analogue to $\psi = \Sigma\, c_n\, \varphi_n\,(x)$ where the function $\psi(x)$ is replaced by a set of coeffs. c_n] are the coefficients ρ_{nm}:

$$\rho(xx') = \sum_{n,m}\rho_{nm}\,\varphi_n(x)\,\bar{\varphi}_m(x')$$

$$= \frac{1}{N}\sum_{\lambda=1}^{N}\rho^{(\lambda)}(xx')$$

$$= \frac{1}{N}\sum_{\lambda=1}^{N}\sum_{n\,m} c_n^{\lambda}\bar{c}_m^{\lambda}\varphi_n(x)\,\bar{\varphi}_m(x') \tag{3-5}$$

hence

$$\rho_{nm} = \frac{1}{N}\sum_{\lambda=1}^{N} c_n^{\lambda}\bar{c}_m^{\lambda} \tag{3-6}$$

in almost all applications we consider the Boltzmann distribution.
Hence.

$$P_n = \rho_{nn} = \frac{1}{Z}\, e^{-\frac{En}{kt}} \tag{3-7}$$

$$\rho_{nm} = 0 \;\; (n \neq m), \qquad \sum P_n = 1 \tag{3-8}$$

If the coefficients are such that they fulfill these conditions at $t = 0$ they will not fulfill them in general at $t \neq 0$ It is one of the assumptions of statistical mechanics that after $t = \tau$, the occurring "phases" will average out again.

A Pure State is one in which no ensemble average is taken (no λ- summation). We show that

$$\rho^2 = \rho \quad \text{in this case:}$$

$$\rho^2 = \int \psi(x)\ \bar{\psi}(x'')\ \psi(x'')\ \bar{\psi}(x')\ dx''$$

$$= \psi(x)\ \bar{\psi}(x') \int \bar{\psi}(x'')\ \psi(x'')\, dx'' \tag{3-9}$$

since we have (silently) assumed that the wave function ψ was normalized.

In the case of a mixed state $\rho^2 < \rho$, this is due to the fact that

$$\left[\frac{1}{Z}\, e^{-\frac{En}{kt}}\right]^2 < \frac{1}{Z}\, e^{-\frac{En}{kt}} \tag{3-10}$$

that is all diagonal elements in this case are < 1.

The pure state corresponds to $T = 0$ (non statistical mechanics) in which

$$P_{\text{ground}} = 1 \qquad P_n = 0 \ (\text{otherwise}) \tag{3-11}$$

and hence

$$\rho^2 = \rho \tag{3-12}$$

(3a) Illustration of "Pure" Versus "Mixed" by Means of Polarization of Light.

Suppose we have a light beam propagated in the z-direction

$$\vec{A}(t,\ z) = \vec{A}\ e^{\,i(\omega t - kz)}\ ;\ \vec{A} = (A_x A_y 0) \tag{3-13}$$

$\theta = tg^{-1}\,(A_y/A_x)$ determines the plane of polarization.

In order to allow, for later considerations, a certain phase factor we assume A_x and A_y to be complex. Both will have the same phase angle α.

The intensity measured with a polarizer that passes light in the plane that makes an angle ϕ with respect to the x-axis is

$$J(\phi) = A_x \bar{A}_x \cos^2\phi + (A_x \bar{A}_y + \bar{A}_x A_y)\ \cos\phi\ \sin\phi$$
$$+ A_y \bar{A}_y \sin^2\phi$$

This can be represented by a Hermitian matrix

$$\rho = \begin{pmatrix} A_x \bar{A}_x & A_x \bar{A}_y \\ A_y \bar{A}_x & A_y \bar{A}_y \end{pmatrix} \tag{3-14}$$

From the preceding it is clear that this corresponds to a pure case: if we measure in the direction $\phi = \theta$ all light will pass. The intensity is $|A^2|$. If we measure in any other direction the intensity will be less: $|A|^2 \cos^2(\phi - \theta)$. If we rotate the x axis such that $\phi = 0$ we have $\vec{A'} = (A'_x, 0, 0)$ this means that the matrix ρ can be transformed into

$$\rho = \begin{pmatrix} 1 0 \\ 0 0 \end{pmatrix} |A|^2 \tag{3-15}$$

and if we take for convenience $|A|^2 = 1 = 1$ we find $\rho^2 = \rho$.

If we superpose two plane polarized waves we find

1) if the relative phase is zero: again a plane polarized wave.
2) if the relative phase is non-zero: an elliptically polarized wave, which may be circularly polarized if special conditions are fulfilled.

In the last case the intensity will not follow a $\cos^2$-law; the density matrix can at most be simplified to

$$\rho = \begin{pmatrix} \alpha\ 0 \\ 0\ 1 - \alpha \end{pmatrix} \tag{3-16}$$

by choosing for the x and y direction the main axis of the ellipse. ($\alpha \neq 0$ and $\alpha \neq 1$) Hence the matrix for the mixed state fulfills

$$\rho^2 \neq \rho \tag{3-17}$$

Another way to make the distinction is to notice that in the first case the matrix was of rank 1, instead of 2 as in the second case.

4 DENSITY OPERATOR

We have:

Density function $\rho(xx')$
Density matrix ρ_{nm}

and we will now introduce Density operator $\hat{\rho}$.

According to the quantum mechanical transformation theory of Dirac, they are identical, i.e., interchangeable.

Introduction of the operator goes as follows:

$$\rho(xx') = \frac{1}{Z} \sum_n \exp(-E_n/kT)\varphi_n(x)\,\overline{\varphi}_n(x'). \tag{4-1}$$

where $\varphi_n(x)$ is the eigenfunction belonging to the Hamiltonian H i. e., the energy operator. Hence we have

$$E_n\varphi_n = H\varphi_n \rightarrow (E_n)^k\varphi_n = (H)^k\varphi_n \tag{4-2}$$

since

$$\sum_k \frac{(-\beta)^k}{k!} E_n^k = e^{-\beta E_n}. \tag{4-3}$$

Hence

$$e^{-\beta E_n}\varphi_n = e^{-\beta H}\varphi_n. \tag{4-4}$$

We find

$$\begin{aligned}\rho(xx') &= \frac{1}{Z}\sum_n \overline{\varphi}_n(x')\,e^{-\beta E_n}\varphi_n(x)\\ &= \frac{1}{Z}\sum_n \overline{\varphi}_n(x')\exp(-\beta H)\varphi_n(x).\end{aligned} \tag{4-5}$$

Introduce the operator

$$\rho_{op} = \frac{1}{Z}e^{-\beta H} = \hat{\rho}$$

If we take the matrix elements of this operator with respect to the energy eigenfunctions.

$$(\hat{\rho})_{mn} = \int \overline{\varphi}_m(x)\,\hat{\rho}\,\varphi_n(x)\,dx \tag{4-6}$$

We recuperate the previous formula for the density function:

$$\sum_{m,n} (\hat{\rho})_{mn}\, \varphi_m(x)\, \overline{\varphi_n}(x')$$
$$= \sum_{mn} \int \varphi_m(x)\, \overline{\varphi_m}(x'')\, \hat{\rho}\, \varphi_n(x'')\, \overline{\varphi_n}(x')\, dx''$$
$$= \sum_{n} \int \delta(x - x'')\hat{\rho}\, \varphi_n(x'')\, \varphi_n(x')\, dx' \tag{4-7}$$
$$= \sum_{n} \frac{1}{Z}\, e^{-En/kt}\quad \varphi_n(x)\, \varphi_n(x') = \rho(xx').$$

If B is an arbitrary operator such as angular momentum or magnetization. We find for the quantum mechanical and statistical expectation value.

$$\ll B \gg\ = \mathrm{Tr}\, (\hat{\rho} B). \tag{4-8}$$

5 WIGNER DISTRIBUTION FUNCTION

It is possible to make a connection between the classical distribution function and the density matrix. This is done in a much more elegant way by Moyal. We will follow the original paper by Wigner.

Definition of the Wigner distribution function is

$$P(x_1 \ldots x_n,\ p_1 \ldots p_n) = \left(\frac{1}{\hbar\pi}\right)^n \int_{-\infty}^{\infty} \ldots \int_{-\infty}^{\infty} dy_1 \ldots dy_n$$
$$e^{2i(p_1 y_1 + \ldots p_n y_n)/\hbar}\ \overline{\psi}\,(x_1 + y_1, \ldots, x_n + y_n)\ \psi\,(x_1 - y_1, \ldots, x_n - x_n) \tag{5-1}$$

We will "collapse" all n variables into one, a vector if one wishes, and in simplified notation:

$$P(x,\ p) = \left(\frac{1}{\hbar\pi}\right)^n \int dy\ e^{\frac{2ipy}{\hbar}}\ \overline{\psi}\,(x + y)\ \psi(x - y) \tag{5-2}$$

Substitute

$$i\hbar \frac{\partial \psi}{\partial t} = -\frac{\hbar^2}{2m} \nabla^2 \psi + V(x)\, \psi(x)$$

$(\frac{\nabla^2}{2m}$ could be replaced by $\sum_{k=1}^{n} \nabla_k^2/2m_k)$

$$-i\hbar \frac{\partial \bar{\psi}}{\partial t} = \frac{\hbar^2}{2m} \nabla^2 \bar{\psi} + V(x)\, \bar{\psi}(x). \tag{5-3}$$

in the first we take $x \to x - y$
in the second we take $x \to x + y$

$$\frac{\partial P}{\partial t} = \left(\frac{1}{\pi\hbar}\right)^n \int dy\, e^{2ipy/\hbar} \left(\frac{\partial \bar{\psi}}{\partial t} \psi + \bar{\psi} \frac{\partial \psi}{\partial t}\right) \tag{5-4}$$

$$\frac{\partial \psi}{\partial t} = \frac{i\hbar}{2m} \nabla^2 \psi - \frac{i}{\hbar} V(x-y)\, \psi(x-y) \tag{5-5a}$$

$$\frac{\partial \bar{\psi}}{\partial t} = -\frac{i\hbar}{2m} \nabla^2 \bar{\psi} + \frac{1}{\hbar} V(x+y)\, \bar{\psi}(x+y) \tag{5-5b}$$

$$\frac{\partial P}{\partial t} = \left(\frac{1}{\pi\hbar}\right)^n \int dy\, e^{2ipy/\hbar} \left\{ -\frac{i\hbar}{2m} [(\nabla^2 \bar{\psi}) \psi - \bar{\psi} \nabla^2 \psi] \right.$$

$$+ \frac{i}{\hbar} V(x+y) \bar{\psi}(x+y) \psi(x-y)$$

$$- \frac{i}{\hbar} V(x-y) \bar{\psi}(x+y)\, \psi(x-y)$$

$$= \text{1st term} + \text{2nd term}. \tag{5-6}$$

1st term

$$\nabla^2 \equiv \frac{\partial^2}{\partial x^2} \text{ is replaced by } \frac{\partial^2}{\partial y^2}$$

$$\frac{\partial}{\partial x} f(x \pm y) = \pm \frac{\partial}{\partial y} f(x \pm y)$$

Then partial integration

$$(\nabla_y^2\,\overline{\psi})\,\psi - \overline{\psi}\;\nabla_y^2\,\psi = \nabla_y\,(\nabla_y\overline{\psi}\cdot\psi - \overline{\psi}\;\nabla_y\psi)$$

$$\int dy\; e^{2ipy/\hbar}\;\{(\nabla_y^2\,\overline{\psi})\,\psi - \overline{\psi}\;\nabla_y^2\,\psi\} =$$

$$= \int dy\; e^{2ipy/\hbar}\;\;\nabla_y\,(\nabla_y\overline{\psi}\,\psi - \overline{\psi}\;\nabla_y\psi)$$

$$= -\int \frac{2ip}{\hbar}\;dy\; e^{2ipy/\hbar}\;(\nabla_y\overline{\psi}\cdot\psi - \overline{\psi}\,\nabla_y\psi) \tag{5-7}$$

replacing ∇_y by ∇_x

$$\left(\text{Note that}\;\begin{matrix}\text{in }\psi: & \nabla_y = -\nabla_x \\ \text{in }\overline{\psi}: & \nabla_y = +\nabla_x\end{matrix}\right) \tag{5-8}$$

$$= -\frac{2ip}{\hbar}\int dy\; e^{2ipy/\hbar}\;(\nabla_x\overline{\psi}\,\psi + \overline{\psi}\;\nabla_x\psi)$$

1st term, including coefficients

$$\left(\frac{1}{\hbar\pi}\right)^n\;\left(-\frac{i\hbar}{2m}\right)\;\left(-\frac{2ip}{\hbar}\right)\int dy\; e^{2ipy/\hbar}\,\nabla_x\,(\overline{\psi}\,\psi) \tag{5-9}$$

$$= -\left(\frac{1}{\hbar\pi}\right)^n\frac{p}{m}\int dy\; e^{2ipy/\hbar}\;\;\nabla_x\,(\overline{\psi}\,\psi) = -\frac{p}{m}\;\nabla\,P(x,\;p).$$

2nd term

$$\left(\frac{1}{\hbar\pi}\right)^n\;\int dy\; e^{2ipy/\hbar}\;\frac{i}{\hbar}(V(x+y) - V(x-y))\,\overline{\psi}\;\psi$$

$$V(x\pm y) = \sum_\lambda\;\frac{(\pm y)}{\lambda!}\;\left(\frac{\partial\;V}{\partial x}\right)_{y=0} \tag{5-10}$$

All even terms cancel

$$V(x+y) - V(x-y) = 2\sum_{\substack{\lambda=1\\ \text{odd}}}\frac{y}{\lambda!}\;\frac{\partial\;V}{\partial x} \tag{5-11}$$

Then the second term becomes

$$\left(\frac{i}{\hbar\pi}\right)^n\;\int dy\; e^{2ipy/\hbar}\,\frac{i}{\hbar}\;2\sum_{\lambda=\text{odd}}\frac{y^\lambda}{\lambda!}\;\;\frac{\partial^\lambda V}{\partial x^\lambda}\;\overline{\psi}\,\psi$$

$$= \left(\frac{1}{\hbar\pi}\right)^n\;\frac{2i}{\hbar}\sum_{\lambda=\text{odd}}\frac{1}{\lambda!}\;\;\frac{\partial\;V}{\partial x}\;\;\int dy\; e^{2ipy/\hbar}\;\;y^\lambda\,\overline{\psi}\,\psi. \tag{5-12}$$

We can "produce" the y^λ by λ-times differentiation of $P(x, p)$ with respect to p, i. e.,

$$\frac{\partial^\lambda P}{\partial p^\lambda} = \left(\frac{1}{\hbar\pi}\right)^n \left(\frac{2i}{\hbar}\right)^\lambda \int dy\; e^{2ipy/\hbar}\; y^\lambda \bar{\psi}\,\psi \qquad (5\text{-}13)$$

hence, 2nd term becomes

$$\left(\frac{1}{\hbar\pi}\right)^n \frac{2i}{\hbar} \sum_{\lambda\,\mathrm{odd}} \frac{1}{\lambda!} \frac{\partial^\lambda V}{\partial x^\lambda} \left(\frac{1}{\hbar\pi}\right)^{-n} \left(\frac{\hbar}{2i}\right) \frac{\partial^\lambda P}{\partial p^\lambda}$$

$$= \sum_{\lambda=\mathrm{odd}} \frac{\partial^\lambda V}{\partial x^\lambda} \frac{(\hbar/2i)^{\lambda-1}}{\lambda!} \frac{\partial^\lambda P}{\partial p^\lambda} \qquad (5\text{-}14)$$

Total result for n - particles:

$$\frac{\partial P}{\partial t} = -\sum_{k=1}^{n} \frac{p_k}{m} \frac{\partial P}{\partial x_k}$$

$$+ \sum_{\lambda_1 + \dots \lambda_n} \frac{\partial^{\lambda_1+\lambda_2 \dots \lambda_n} V}{\partial x_1^{\lambda_1}\, \partial x_2^{\lambda_2} \dots \partial x_n^{\lambda_n}}$$

$$\frac{\left(\frac{\hbar^2}{2i}\right)^{\lambda_1 + \dots \lambda_n - 1}}{\lambda_1!\lambda_2! \dots \lambda_n!} \frac{\partial^{\lambda_1 + \dots \lambda_n} P}{\partial^{\lambda_1} p_1 \dots \partial^{\lambda_n} p_n}$$

second term sum over

$$\lambda_1,\ \lambda_2 \dots \lambda_n \ \text{ with } \lambda_1 + \lambda_2 + \lambda_3 \dots \lambda_n = \text{odd} \qquad (5\text{-}15)$$

For harmonic oscillator, identical with Boltzmann transport equation. (Note the fact that the diff. Eq. is the same does not imply that the solution is the same.)

6 DISCUSSION WIGNER DISTRIBUTION FUNCTION.

According to the Born Interpretation we have

$$\rho(x, x)\; dx = \psi(x)\, \bar{\psi}(x)\; dx = P(x)\; dx \qquad (6\text{-}1)$$

in the same way, after a Fourier transform:

$$\psi(x) = \left(\frac{1}{2\pi\hbar}\right)^{n/2} \int \phi\,(p)\, e^{-\frac{ipx}{\hbar}}\, dp \qquad (6\text{-}2)$$

and we can write down again that the density matrix in momentum space has diagonal elements which are the probability for finding a certain value of the momentum

$$\rho(p,\ p)\ dp = \phi(p)\ \overline{\phi}(p)\, dp = P(p)\, dp. \qquad (6\text{-}3)$$

that is, a momentum between p and p + dp.

We will show below that the Wigner function will give both:

$$1)\ \int P_w(x,\ p)\ dp = P(x)\,;$$

$$2)\ \int P_w(x,\ p)\, dx = P(p) \qquad (6\text{-}4)$$

The first is straight forward and is based on the definition of the deltafunction:

$$\int P_w(x,\ p)\, dp = \left(\frac{1}{\pi\hbar}\right)^n \int dy\ e^{\frac{2ipy}{\hbar}}\ \overline{\psi}(x+y)\,\psi(x-y)\, dp =$$

$$= \left(\frac{1}{\pi\hbar}\right)^n \int dy\, \overline{\psi}(x+y)\,\psi(x-y) \int dp\, e^{2ipy/\hbar} \qquad (6\text{-}5)$$

$$\int e^{\frac{2ipy}{\hbar}}\ dp = \left(\frac{\hbar}{2}\right)^n (2\pi)^n\ \delta(y) = (\pi\hbar)^n\ \delta(y) \qquad (6\text{-}6)$$

which gives 1)

For the second we write the factor 2y in the definition of the Wigner function as: 2y = (x + y) - (x - y). Then new integration variables are introduced and the expression becomes the product of two Fourier transforms:

$$\int P_w(y,\ p)\, dx = \left(\frac{1}{\pi\hbar}\right)^n \int dy\, \overline{\psi}(x+y)\, e^{\frac{ip(x+y)}{\hbar}}\, \psi(x-y)\, e^{-\frac{ip(x-y)}{\hbar}} dx \qquad (6\text{-}7)$$

$$u = x + y \qquad \text{Jacobian } \left(\frac{1}{2}\right)^n \text{ hence}$$
$$v = x - y$$

$$\int P_w(x,\ p)\, dx = \left(\frac{1}{2\pi\hbar}\right)^n \int du\, \overline{\psi}(u)\, e^{\frac{ipu}{\hbar}} \int dv\, \psi(v)\, e^{-\frac{ipv}{\hbar}} =$$

$$= \phi(p)\ \overline{\phi}(p) \qquad (6\text{-}8)$$

It is easy to show that $\overline{P}_w = P_w$ by replacing y by -y after complex conjugation the formula reproduces itself. Hence P_w is a real function. It is however not always positive:

As an example of a negative P_w we take the first excited state of an harmonic oscillator:

$$\psi(x) = x\,e^{-\frac{1}{2}x^2} \to \overline{\psi}(x+y)\,\psi(x+y) = (x^2 - y^2)\,e^{-(x^2+y^2)}$$

$$P_w(x, k) = x^2\, e^{-x^2} \int e^{2iky}\, e^{-y^2}\, dy - e^{-x^2} \int y^2\, e^{2iky}\, e^{-y^2}\, dy$$

$$\text{if } k = 0 \qquad P_w(x, 0) = x^2 e^{-x^2}\sqrt{\pi} - \frac{1}{2}\, e^{-x^2}\sqrt{\pi} \tag{6-10}$$

$$= \left(x^2 - \frac{1}{2}\right) e^{-x^2}\sqrt{\pi} \text{ is ng. for } \left| \frac{-1}{\sqrt{2}} < x < \frac{1}{\sqrt{2}} \right.$$

Note: It is possible to determine the Wigner distribution function for the oscillator function with arbitrary n (a Laguerre polynomial of order n multiplied by an exponential).

The first two properties of the Wigner function are similar to the properties of the classical distribution function. The fact that the W. D. f. can be negative excludes the possibility of interpreting it as a probability function however. This is to be expected on the ground of the uncertainty relations.

7 EXAMPLES OF THE DENSITY MATRIX

This is the canonical density matrix:

1) Free particles

$$\rho(xx') = \frac{1}{Z} \int e^{-\beta\frac{\hbar^2 k^2}{2m}}\, e^{-ikx'}\, e^{ikx}\, dx. \tag{7-1}$$

2) Particles in a box (Length one: $0 < x < 1$) $\quad h = m = 1$

$$\rho(xx') = \frac{1}{Z} \sum_{n=1}^{\infty} \left(\exp. -\frac{\beta\pi^2 n^2}{2}\right) 2\sin(\pi n x)\sin(\pi n x'). \tag{7-2}$$

This expression can be obtained in a more or less closed form; see discussion Bloch equation.

3) Harmonic Oscillator.

$$\rho(xx') = \frac{1}{Z}\sum_{n=0}^{\infty} e^{-\beta E_n} N_n^2 H_n(x) H_n(x') e^{-\frac{1}{2}(x^2+x'^2)}$$

$$E_n = \hbar\omega(n + \frac{1}{2}) \qquad N_n^{-2} = \sqrt{\pi}\, 2^n n! \qquad (7\text{-}3)$$

This sum can be expressed in closed form. The essential part of it we call F(x, y; t) and is evaluated below. (This formula is also well known in mathematical statistics. A completely different application is in the papers of Kac and Uhlenbeck where it is used to determine the partition function of a special model-gas.)

$$F(x, y; t) = \frac{1}{\sqrt{\pi}} \sum_{n=0}^{\infty} \frac{(t/2)^n}{n!} H_n(x) H_n(y) e^{-\frac{1}{2}(x^2+y^2)} \qquad (7\text{-}4)$$

$$\int F(x, y; t) H_n(y) \exp(-\frac{1}{2}y^2)\, dy$$

$$= \frac{1}{\sqrt{\pi}} \frac{(t/2)^n}{n!} H_n(x) e^{-\frac{1}{2}x^2} \sqrt{\pi}\, 2^n n! \qquad (7\text{-}5)$$

$$= t^n H_n(x) e^{-\frac{1}{2}x^2}$$

After multiplying both sides with the Hermite function of the argument y, using orthogonality, multiplying with $s^n/n!$ and summing, we obtain:

$$\int F(x, y; t) e^{-s^2+2sy} e^{-\frac{1}{2}y^2} dy = e^{-(st)^2+2stx} e^{-\frac{1}{2}x^2} \qquad (7\text{-}6)$$

The integral transforms a Gaussian into a Gaussian, hence the kernel has to be a Gaussian also. We assume arbitrary coefficients, which in themselves are functions of the parameter t. We then perform the integral, which is just a question of completing of the squares and compare the coefficients of x^2, x and the constant with the same coefficients on the right-hand side.

$$F(x, y) = c\, e^{ax^2+by^2+2cxy} \qquad (7\text{-}7)$$

The exponent is:

$$(b - \frac{1}{2})\, y^2 + 2(s + cx)y + ax^2 - s^2$$

$$= (b - \frac{1}{2}) \left\{ y^2 + 2\,\frac{s + c}{b - \frac{1}{2}}\, y + \frac{s + cx}{b - \frac{1}{2}}^{\,2} \right\} - \frac{(s + cx)^2}{b - \frac{1}{2}}$$

$$+ ax^2 - s^2 \tag{7-8}$$

Hence the equations that determine the coefficients a, b, c are:

$$a - \frac{c^2}{b - \frac{1}{2}} = -\frac{1}{2}; \quad -\frac{2sc}{b - \frac{1}{2}} = 2st; \quad \frac{-s^2}{b - \frac{1}{2}} - s^2 = -s^2 t^2$$

with the result: $b = -\frac{1}{2}\left(\frac{1 + t^2}{1 - t^2}\right)$ Since $t < 1$ the leading coefficient is negative.

$$c = \frac{t}{1 - t^2} \qquad a = b$$

$$ax^2 + by^2 + 2cxy = -\frac{(1 + t^2)(x^2 + y^2) - 4txy}{2(1 - t^2)}$$

$$= -\frac{(1 - t^2)(x + y)^2 + (1 + t)^2 (x - y)^2}{4(1 - t^2)}$$

$$= \frac{1}{4}\left\{\frac{1 - t}{1 + t}(x + y)^2 + \frac{1 + t}{1 - t}(x - y)^2\right\}; \quad t = e^{-\beta\hbar\omega} \tag{7-9}$$

$$\frac{1 - t}{1 + t} = \frac{1 - e^{-\beta\hbar\omega}}{1 + e^{-\beta\hbar\omega}} = \frac{e^{\frac{\beta\hbar\omega}{2}} - e^{-\frac{\beta\hbar\omega}{2}}}{e^{\frac{\beta\hbar\omega}{2}} + e^{-\frac{\beta\hbar\omega}{2}}} = \text{tgh}\,\frac{\beta\hbar\omega}{2} \tag{7-10}$$

The factor in front of the exponential of F is:

$$\frac{1}{C} = \sqrt{\frac{\pi}{\frac{1}{2} - b}} = \sqrt{\pi}(1 - t^2) \tag{7-11}$$

$$F(x,\ y;t) = \frac{1}{\sqrt{\pi(1-t^2)}} \exp\left[-\frac{1}{4}\left\{\text{tgh}\left(\frac{\beta\hbar\omega}{2}\right).(x+y)^2 + \text{cotgh}(\ ).(x-y)^2\right\}\right] \quad (7\text{-}12)$$

hence the density function is:

$$\rho(xx') = \frac{1}{Z} e^{-\frac{\hbar\omega}{2kT}} F(x,\ x';t) = \frac{1}{Z} \frac{t}{\sqrt{\pi(1-t^2)}} e^{[\ldots]}$$

$$= \frac{1}{Z} \frac{e^{[\ldots]}}{\sqrt{2\pi\ \sinh\ \beta\hbar\omega}} \quad (7\text{-}13)$$

To determine Z we take the trace of the density function:

$$1 = \int \rho(x,\ x)\ dx = \frac{1}{Z} \frac{\sqrt{\frac{\pi}{\text{tgh}(\beta\hbar\omega/2)}}}{\sqrt{2\pi\ \sin\beta\hbar\omega}} =$$

$$\int e^{-\frac{1}{4}\alpha(2x)^2} dx = \sqrt{\frac{\pi}{\alpha}} \qquad \frac{\sqrt{\frac{e^{x/2} + e^{-x/2}}{e^{x/2} - e^{-x/2}}}}{\sqrt{(e^{x/2} - e^{-x/2})(e^{x/2} + e^{-x/2})}} = \frac{e^{\beta\hbar\omega}}{e^{\beta\hbar\omega} - 1}$$

$$\alpha = \text{tgh}(\beta\hbar\omega/2) \quad (7\text{-}14)$$

this result could of course have been obtained much quicker by the summation of a geometric series

$$\text{tr}\ \rho(xx) = \text{tr} \sum_{n=0}^{\infty} e^{-\beta\hbar\omega(n+\frac{1}{2})}\ \overline{\psi}_n(x)\ \psi_n(x)$$

$$= \sum_{n=0}^{\infty} e^{-\beta\hbar\omega(n+\frac{1}{2})} \underbrace{\text{tr}\ \psi_n(x)\ \psi_n(x)}_{1} \quad (7\text{-}15)$$

It is possible to obtain the Wigner distribution function without any trouble. One term will fall outside the integral, the other is a Gaussian integral

The result is

$$P_w = \frac{1}{2\pi\cosh(\hbar\omega\beta/2)} e^{-\frac{1}{4}\text{tgh}\left(\frac{\hbar\beta}{2}\right)(p^2 + x^2)} \quad (7\text{-}16)$$

(the classical limit $\hbar \to 0$ gives $\frac{1}{2\pi} e^{-\beta/2(p^2+x^2)}$)

Note. An alternate way to calculate P_w for the canonical ensemble is to construct first the P_w's for each individual Hermite function. These P_w's are the functions belonging to pure density matrices. A linear combination of these pure density matrices with the appropriate Boltzman factors is the canonical (mixed) density matrix, and that same statement holds for the Wigner distribution functions. Hence if we take the P_w's of each Hermite function and sum these with the Boltzman factors in front, we will obtain the formula above. It turns out that the mathematics involved is equivalent to the construction of a generating function for the Laguerre polynomials.

$$(P_w)_{can} = \frac{1}{N} \sum P_\lambda P_w^{(\lambda)} \tag{7-17}$$

8 BLOCH EQUATION.

$$\text{From } \rho(xx') = \sum_{n=0}^{\infty} e^{-\beta E_n} \psi_n(x) \bar{\psi}_n(x') \tag{8-1}$$

$$\text{we find } \frac{\partial \rho}{\partial \beta} = - \mathcal{H}_{op} \rho \tag{8-2}$$

this differential equation will give back the density function provided we invoke the initial condition:

$$\lim_{\beta \to +0} \rho = \delta(x - x') \tag{8-3}$$

This equation is nearly the same as the Schrödinger equation. Hence all techniques which are used for the Schrödinger equation can be used here, too. The only difference is the interval of the variable: $0 < \beta < \infty$ but $-\infty < t < \infty$. The time equation contains "i," the temperature equation does not; i. e., the time has been oscillatory part, the temperature is monotonous.

Note: ρ is not normalized to $\operatorname{tr} \rho = 1$, i. e., the factor $1/Z$ is left out (for normalized ρ the differential equation will have an additional term).

Example 1: Free particles

$$\rho(xx') = \int e^{-\frac{\hbar^2 k^2}{2m}\beta} e^{ikx} e^{-ikx'} dk. \tag{8-4}$$

The integral is of the form:

$$\int e^{-ax^2} e^{ibx} dx = \frac{\pi}{a} e^{\frac{-b^2}{4a}} \tag{8-5}$$

hence the density function for a free particle is:

$$\rho(xx') = \sqrt{\frac{2\pi m}{\beta\hbar^2}}\ e^{-\frac{m(x-x')^2}{2\hbar^2\beta}} \tag{8-6}$$

The factor in front of the exponential is the well known "thermal-deBroglie wavelength." Note that the integral is suited for a β expansion (high-temperature expansion) while the answer is suited for a low-temperature expansion.

The corresponding Bloch equation is

$$\frac{\partial\rho}{\partial\beta} = \frac{m(x-x')^2}{2\hbar^2\beta^2}\ e^{-\frac{m(x-x')^2}{2\hbar^2\beta}} = -\mathcal{H}\rho = \frac{\hbar^2}{2m}\nabla^2\rho. \tag{8-7}$$

This differential equation is the same as the diffusion equation as will be discussed below.

Example 2: Free particle in a box.

We take $\hbar = m = 1$ and the interval $0 < x < 1$ hence the boundary condition $\psi(0) = \psi(1) = 0$.
The normalized wave functions are $\psi_k = \sqrt{2}\ \sin(k\pi x)$ and the energy $E = \frac{1}{2}k^2$ the density function

$$\rho\ (xx') = \sum_{k=1}^{\infty} e^{-\beta\frac{n^2k^2}{2}}\, 2\sin(k\pi x)\sin(k\pi x')$$

If we start with the Bloch equation we can obtain an entirely different looking equation. The Bloch equation is, as mentioned above, the same as the heat conduction equation (alias the diffusion equation, alias the Brownian motion equation) which is:

$$\frac{\partial}{\partial \beta}\rho = c\,\nabla^2\,\rho \qquad\qquad c = \hbar^2/2m. \tag{8-8}$$

it has the solution

$$\rho = \frac{1}{\sqrt{4c\beta}}\, e^{-\frac{(x-x')^2}{4c\beta}} \tag{8-9}$$

which represents a Gaussian of constant area which has an increasing width for growing β. At $\beta = 0$ it is a delta function. Of course any superposition is also a solution. The superposition can be discrete or with the help of an integral (in the last case the expression above can be looked upon as the Green's function of the problem).

We check the solution by substitution:

$$\frac{\partial \rho}{\partial \beta} = -\frac{1}{2\beta}\,\rho + \frac{(x-x')^2}{4c\beta^2}\,\rho \tag{8-10}$$

$$\frac{\partial \rho}{\partial x} = \frac{-2(x-x')}{4c\beta}\,\rho \tag{8-11}$$

$$\frac{\partial^2 \rho}{\partial x^2} = -\frac{1}{2c\beta}\,\rho - \frac{(x-x')}{2c\beta}\,\rho' = \frac{1}{c}\quad -\frac{1}{2\beta}\,\rho + \frac{2(x-x')^2}{8c\beta^2}\,\rho \tag{8-12}$$

hence it fulfills the differential equation. It also fulfills the initial condition. But it does not satisfy the boundary condition. This can be done by superposition. Method of the images (Sommerfeld, differential equations).

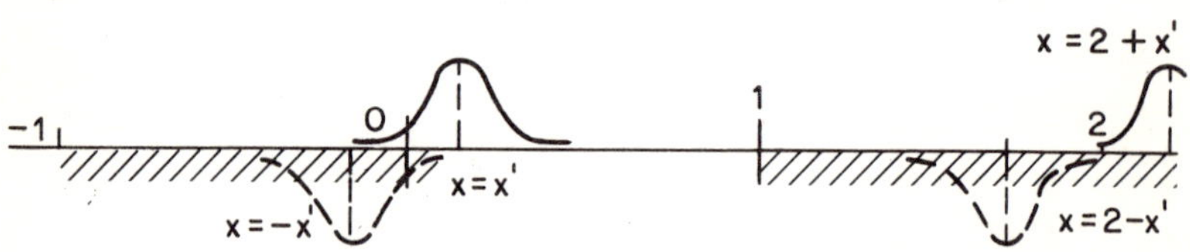

Instead of one Gaussian we repeat them at distances 2n to make the problem periodic. Also we want the points 0 and 1 to be nodes. In order to accomplish this we make it antisymmetric around these points. This is done by putting an upside-down Gaussian on the points -x' and 2 - x' and so on. The total result is:

$$\rho(xx') = \frac{1}{\sqrt{2\pi\beta}} \sum_{n=\infty}^{\infty} \left(e^{-\frac{(x'-x+2n)^2}{2\beta}} - e^{-\frac{(x'+x+2n)^2}{2\beta}}\right) \tag{8-13}$$

This is a low temperature expansion. (The two different sums are two different representations of an elliptic function.)

Example 3: The Harmonic Oscillator. For the density matrix of the harmonic oscillator see previous section.

Problem: calculate the Wigner distribution function, for example 1 and 2.

9. SOLUTION OF THE BLOCH EQUATION BY SUCCESSIVE APPROXIMATIONS.

If we have $\frac{\partial \rho}{\partial \beta} = -\mathcal{H}\rho \qquad \mathcal{H} = \mathcal{H}_0 + \lambda\ \mathcal{H}_1 \qquad$ (9-1)

and if the eigenvalues, eigenfunctions, density matrix, etc., <u>are known</u> for $\mathcal{H}_0$ (usually $\mathcal{H}_0$ is the kinetic energy, hence plane waves) what can be said about $\mathcal{H}_{\mathbf{total}}$?

$$\mathcal{H}_1(u) = e^{u\mathcal{H}_0}\mathcal{H}_1 e^{-u\mathcal{H}_0}$$

The operator ρ fulfills the so called "Dyson-equation"

$$\rho(\beta) = e^{-\beta\mathcal{H}_0}\left[1 - \lambda \int_0^{\beta} \mathcal{H}_1(u)\ e^{u\mathcal{H}_0}\rho(u)\,du\right] \tag{9-2}$$

This equation is much older and can be found in matrix form in Born and Jordan. This equation is the starting point for many iteration methods. It is the integration equation equivalent of the Bloch differential equation. It has as such the advantage that the initial condition is built in.

The simplest way to prove it is to introduce the interaction representation:

$$\frac{\partial \rho}{\partial \beta} = -\mathcal{H}_0\rho - \lambda\mathcal{H}_1\rho \qquad \rho = e^{-\beta\mathcal{H}_0}S \tag{9-3}$$

$$\frac{\partial \rho}{\partial \beta} = - \mathcal{H}_0 \ \rho + e^{-\beta\mathcal{H}_0} \frac{\partial S}{\partial \beta} \tag{9-4}$$

$$e^{-\beta\mathcal{H}_0} \frac{\partial S}{\partial \beta} = - \lambda \mathcal{H}_1 \rho \longrightarrow \frac{\partial S}{\partial \beta} = - \lambda \, e^{+\beta\mathcal{H}_0} \mathcal{H}_1 \, e^{-\beta\mathcal{H}_0} S \tag{9-5}$$

$$\frac{\partial S}{\partial \beta} = - \lambda \mathcal{H}_1 (\beta) S \quad \text{with } \mathcal{H}_1(\beta) = e^{\beta\mathcal{H}_0} \ \mathcal{H}_1 \ e^{-\beta\mathcal{H}_0} \tag{9-6}$$

The solution of this equation is:

$$S = -\lambda \int_0^\beta \mathcal{H}_1 (u) \ S(u) \, du + C \tag{9-7}$$

with the initial condition $\beta = 0 \rightarrow S = 1 \rightarrow C = 1$, hence

$$\rho(\beta) = e^{-\beta\mathcal{H}_0} \left[1 - \lambda \int_0^\beta \mathcal{H}_1 (u) \, e^{u\mathcal{H}_0} \rho(u) \, du\right] \tag{9-8}$$

We will now give another derivation of this crucial equation, using the Green's function, because it shows the similarity with the resolvent method to be discussed next.

We consider the right-hand side as a "driving term," that is we consider the equation as an inhomogeneous equation in which the last term is supposedly known.

$$\left(\frac{\partial}{\partial \beta} + \mathcal{H}\right) \rho = 0 \rightarrow \left(\frac{\partial}{\partial \beta} + \mathcal{H}_0\right) \rho = - \lambda \ \mathcal{H}_1 \ \rho . \tag{9-9}$$

In order to determine the particular solution of this equation we determine the Green's function $K(\beta, \beta')$ from:

$$\left(\frac{\partial}{\partial \beta} + \mathcal{H}_0\right) K(\beta\beta') = \delta(\beta - \beta') . \tag{9-10}$$

Take the Fourier transform of K

$$K(\beta\beta') = \int S(\epsilon) \ e^{-i\epsilon(\beta-\beta')} \quad d\epsilon \tag{9-11}$$

Differentiation with respect to β gives a factor $-i\epsilon$ in the integral and the δ-function gives a factor $1/2\pi$:

$$\frac{\partial K}{\partial \beta} = - i \int \epsilon \; S(\epsilon)\, e^{-i\epsilon(\beta-\beta')} \; d\epsilon \tag{9-12}$$

$$\delta(\beta - \beta') = \frac{1}{2\pi} \int e^{-i\epsilon(\beta-\beta')} \; d\epsilon \tag{9-13}$$

Hence
$$(\mathcal{H}^0 - i\epsilon)\, S(\epsilon) = \frac{1}{2\pi} \tag{9-14}$$

and
$$S(\epsilon) = \frac{1/2\pi}{\mathcal{H}^0 - i\epsilon} \tag{9-15}$$

The result for K is

$$K(\beta\beta') = \frac{1}{2\pi} \int \frac{e^{-i\epsilon(\beta-\beta')}}{\mathcal{H}^0 - i\epsilon} \; d\epsilon = \frac{1}{2\pi i} \int \frac{e^{-E(\beta-\beta')}}{\mathcal{H}^0 - E} \; dE \tag{9-16}$$

the last term is obtained after a substitution $i\epsilon = E$ and the tacit assumption is made that the function can be analytically continued in all appropriate regions.

The Cauchy theorem gives:

$$K(\beta\beta') = e^{-\mathcal{H}_0(\beta-\beta')} \tag{9-17}$$

Actually, it was never proved for operators, but one can argue that if it holds for discrete eigenvalues, it will also hold for operators, and conversely.

The final result is:

$$\rho = \rho_0 - \lambda \int_0^\beta e^{-\mathcal{H}_0(\beta-\beta')} \; \mathcal{H}_1 \rho. \tag{9-18}$$

the first term is the general solution with a constant (one) in accordance with the initial condition. The answer can be written in the same form as the interaction representation method gave.

10. RESOLVENT METHOD.

The last method is connected with the resolvent method (called Kubo's method in statistical mechanics and van Hove's method in field theory). The density operator (its time equivalent is sometimes called the propagator since it transforms a wavefunction at time t into one at time t').

$$\psi(t') = e^{-i\mathcal{H}(t-t')}\psi(t') \qquad (10\text{-}1)$$

in each case: if we take the exponential operator we can write it as a Cauchy integral. (For continuous eigenvalues this is somewhat dubious mathematics but this is never emphasised too strongly.)

$$\rho(\beta) = e^{-\beta\mathcal{H}} = \frac{1}{2\pi i} \oint \frac{e^{-\beta E}}{E - \mathcal{H}}\, dE \qquad (10\text{-}2)$$

since

$$\left(\frac{1}{E - \mathcal{H}}\right)_{op} \psi_n = \frac{1}{E - E_n}\, \psi_n \rightarrow \text{residue } e^{-\beta E_n} \qquad (10\text{-}3)$$

In this case $\mathcal{H}$ was the total hamiltonian, contrary to the previous use of the Cauchy integral. Again separating in a main and a perturbation part, we have:

$$\frac{1}{E - \mathcal{H}} = \frac{1}{E-(\mathcal{H}_0 + \lambda\, \mathcal{H}_1)} \qquad (10\text{-}4)$$

this can be powerseries-developed in a geometrical series, the difference being that the small part is "sandwiched" between the denominators in order to preserve the proper order of the operators:

$$\frac{1}{E - \mathcal{H}} = \frac{1}{E - \mathcal{H}_0} + \frac{1}{E - \mathcal{H}_0}\lambda\mathcal{H}_1 \frac{1}{E - \mathcal{H}_0}$$

$$+ \frac{1}{E - \mathcal{H}_0}\lambda\mathcal{H}_1 \frac{1}{E - \mathcal{H}_0}\lambda\mathcal{H}_1 \frac{1}{E - \mathcal{H}_0}$$

$$+ \frac{1}{E - \mathcal{H}_0}\lambda\mathcal{H}_1 \frac{1}{E - \mathcal{H}_0}\lambda\mathcal{H}_1 \frac{1}{E - \mathcal{H}_0}\lambda\mathcal{H}_1 \frac{1}{E - \mathcal{H}_0}$$

$$+ \dots \qquad (10\text{-}5)$$

This formula is proved by multiplication with the denominator:

$$(E - \mathcal{H}_0 - \lambda\mathcal{H}_1)\frac{1}{E - \mathcal{H}_0 - \lambda\mathcal{H}_1} = (E - \mathcal{H}_0 - \lambda\mathcal{H}_1)\frac{1}{E - \mathcal{H}_0}$$

$$+ (E - \mathcal{H}_0 - \lambda\mathcal{H}_1)\frac{1}{E + \mathcal{H}_0}\lambda\mathcal{H}_1\frac{1}{E - \mathcal{H}_0}$$

$$+ (E - \mathcal{H}_0 - \lambda\mathcal{H}_1)\frac{1}{E - \mathcal{H}_0}\lambda\mathcal{H}_1\frac{1}{E - \mathcal{H}_0}\lambda\mathcal{H}_1\frac{1}{E - \mathcal{H}_0}$$

$$= (E - \mathcal{H}_0)\frac{1}{E - \mathcal{H}_0} - \lambda\mathcal{H}_1\frac{1}{E - \mathcal{H}_0} + (E - \mathcal{H}_0)\frac{1}{E - \mathcal{H}_0}\lambda\mathcal{H}_1\frac{1}{E - \mathcal{H}_0}$$

$$- \lambda\mathcal{H}_1\frac{1}{E - \mathcal{H}_0}\lambda\mathcal{H}_1\frac{1}{E - \mathcal{H}_0}$$

$$+ (E - \mathcal{H}_0)\frac{1}{E - \mathcal{H}_0}\lambda\mathcal{H}_1\frac{1}{E - \mathcal{H}_0}\lambda\mathcal{H}_1\frac{1}{E - \mathcal{H}_0} + \ldots = 1 \tag{10-6}$$

A slightly less pedestrian way is to start with the operator tautology

$$\frac{1}{E - \mathcal{H}_0 - \mathcal{H}_1} = \frac{1}{E - \mathcal{H}_0}\left[1 + \left(\mathcal{H}_1 + \mathcal{H}_1\frac{1}{E - \mathcal{H}_0 - \mathcal{H}_1}\mathcal{H}_1\right)\frac{1}{E - \mathcal{H}_0}\right] \tag{10-7}$$

and to iterate this expression.

Note: there is a relation between this and the summing up of the most divergent diagrams.

11 ITERATION AND TIME ORDERING.

We introduce for $\lambda\mathcal{H}_1$ the interaction potential $V(r' - r)$ which is not an operator in the narrow sense of the word, but in second quantisation it will be one. Hence, it is not only not commuting with the kinetic energy, but also with itself (at a different "time"). We have to keep this in mind during the evaluation of the Boltzmann factor. We separate the main part of the Boltzmann exponential:

$$S(\beta) = 1 - \int_0^\beta V(u)\; S(u)\, du \quad \text{with } S(0) = 1 \tag{11-1}$$

1st approximation:

$$S^{(1)}(\beta) = 1 - \int_0^\beta du\; V(u) \tag{11-2}$$

2nd approximation:

$$\begin{aligned} S^{(2)}(\beta) &= 1 - \int_0^\beta du\, V(u)\; S^{(1)}(u)\, du \\ &= 1 - \int_0^\beta du\; V(u) + \int_0^\beta du \int_0^u du'\; V(u)\; V(u') \\ &= S_0 + S_1 + S_2 \end{aligned} \tag{11-3}$$

(the order of V and S has to be kept!)

$$S_0 = 1 \qquad\qquad S_1 = -\int_0^\beta du\; V(u) \tag{11-4}$$

$$S_2 = + \int_0^\beta du_2 \int_0^{u_2} du_1\; V(u_2)\; V(u_1) \qquad \begin{cases} u = u_2 \\ u' = u_1 \end{cases} \tag{11-5}$$

$0 < u_1 < u_2$ that is u_2 is always "later" (larger) than u_1.

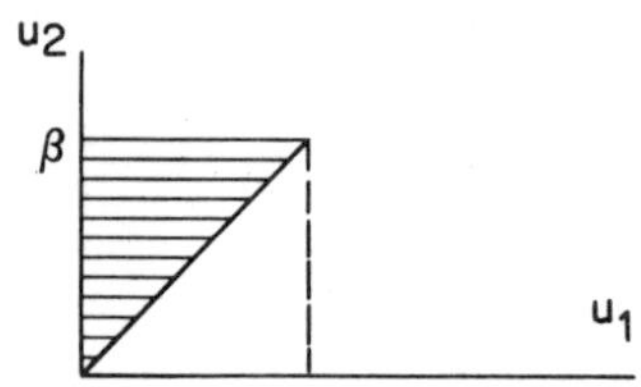

range of integration

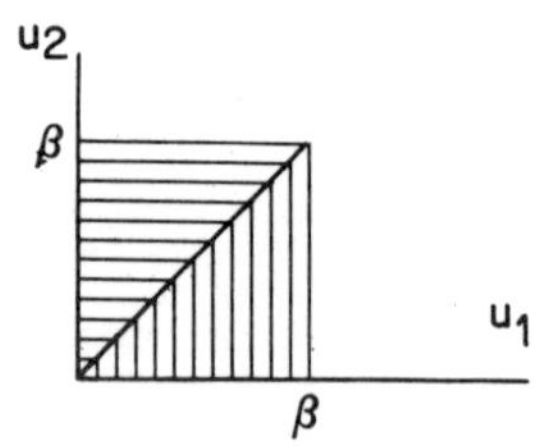

range after time ordering has been introduced.

Integral is taken over a square by adding second triangle: The meaning of the symbol T (time ordering operator) is:

if u_2 is smaller that u_1 we have to interchange the V's:

if $u_2 < u_1$ replace $V(u_2)V(u_1)$ by $V(u_1)V(u_2)$

in which case (if we call $u_2 \to u_1$ and $u_1 \to u_2$) we have the original integral back (with the condition that $u_1 < u_2$ as it should be) and hence we are allowed to integrate over the square provided we introduce a factor $\frac{1}{2}$.

In general we have n! ways of writing n factors V's, but only one of them is correct (i. e., in descending order) hence we need 1/n!

$$S = \sum_{n=0}^{\infty} \frac{(-1)^n}{n!} \, T \int_0 du_n \int_0 du_{n-1} \int_0 \cdots \prod_{i=1}^{n} V(u_i).$$

$$S = T\, e^{-\int_0^\beta V(u')\,du'} \tag{11-6}$$

Undo the interaction representation:

$$\rho = e^{-\beta H_0} S(\beta). \tag{11-7}$$

$$Z = \mathrm{tr}\,\rho = \mathrm{tr}\left\{e^{-\beta H_0} S(\beta)\right\} \tag{11-8}$$

$$\frac{Z}{Z_0} = \frac{\mathrm{Tr}\left\{e^{-\beta H_0} S(\beta)\right\}}{\mathrm{Tr}\left\{e^{-\beta H_0}\right\}} \tag{11-9}$$

Average over arbitrary physical quantity A:

$$\langle A \rangle = \frac{\mathrm{Tr}\left\{e^{-\beta \mathcal{H}_0} A\right\}}{\mathrm{Tr}\left\{e^{-\beta \mathcal{H}_0}\right\}} \tag{11-10}$$

we use this bracket notation:

$$\frac{Z}{Z_0} = \sum_{n=0}^{\infty} \frac{(-1)^n}{n} \int_0 \cdots \int_0 \prod_{k=1}^{n} du_k \langle T\, V(u_k)\rangle. \tag{11-11}$$

12 SHORT DIGRESSION ON SECOND QUANTIZATION

Introduce an abstract wave function which does nothing but counting the number of "particles." The intensity of the wave (amplitude square) is proportional to the number of "particles."

Let us call it a string for convenience and also to stress the fact that it has nothing to do with an ordinary wave function.

We describe this string with the quantum mechanics of the harmonic oscillator: let Ψ_0 be the n = 0 ground state.

Ψ_0 = "vacuum" (no particles)

$a_k^+ \Psi_0$ = let there be one particle i. e., wave function (with energy E_k)

$a_k^+ a_{k'}^+ \Psi_0$ = let there be two wave functions one with energy $E_{k'}$ another with E_k.

Arguments in favor of the consistency of this picture.

1. A quantum mechanical string is quantized, hence the intensity are multiples of a unit intensity. Hence we can use this device to indicate the number of particles.
2. We usually describe the situation by stating the energy (or other relevant dynamic variables) of particle one, of particle two and so on.

Already in classical statistical mechanics the "generic" way of counting is used, that is, one indicates how many particles between ν and $\nu + d\nu$ instead of specifying the velocity of each particle (specific).

3. The Number operator:

$a_k^n \Psi_0 \equiv$ (there are n particles)

the operator $a_k^+ a_k$ has the eigenvalue n.

$$\underbrace{a_k^+ a_k}_{\text{operator}} (a_k^n \Psi_0) = \underbrace{n_k}_{\text{eigenvalue}} \Psi_0 .$$

Hence, energy $E = \Sigma\, n_k \epsilon_k$ can be written as

$$E_{op} = \sum \epsilon_k a_k^+ a_k . \qquad (12\text{-}1)$$

which is an operator in sense of second quantization.

Note: second quantization does not introduce h.

Advantage: (which is not exploited in statistical mechanics) is the variable number of particles; beta-decay can be described with this formalism.

Final goal is to write the potential energy also in this way.

The ideas sketched above enable us only to translate a diagonal operator (like the kinetic energy with respect to a set of plane wave basis functions) into its second quantized equivalent.

In order to do the same with a nondiagonal operator Q, that is an operator which has, with respect to a certain set of orthonormal wave functions $\{\psi_m\}$, elements $Q_{mm'}$ that are non zero. We will restrict our discussion to bosons. (No exclusion principle.)

A many particle wave function is (or better: can be written as) a linear combination of a product (set of products) of one particle eigenfunctions

$$\psi_{m_1}(x_1)\ \psi_{m_2}(x_2)\ \ldots\ \psi_{m_n}(x_N) = \Pi\ \psi_{m_i}(x_i) \qquad (12\text{-}2)$$

total symmetrization is necessary for bosons, hence we replace this by

$$\sum^{N} P\ \psi_{m_1}(x_1) = \psi_{\substack{\text{many particle}\\ \text{symmetrized}}} \qquad (12\text{-}3)$$

after which we do not have to worry about the symmetry any more since all operators are symmetric hence will turn a symmetric wave function into a symmetric one.

The meaning of P:

a) The permutation can be performed either on the particles (coordinates) or on the labels (quantum number).

One permutation is the inverse of the other but after summation this is irrelevant

b) Tolman, Condon, and Shortley and many others perform all N! permutations.

c) Landau -Lifshitz consider permutation on labels, and only on labels that have different values. This simplifies matters somewhat, and is moreover in accordance with the true spirit of quantum mechanics that "particles" have no identity and are only recognizable by the state in which they are. The total number of terms in the sum will be

$$\frac{N!}{N_1!\,N_2!\,N_3!\ldots} \quad \text{instead of } N!$$

Consequently it is much handier to label the ψ generically instead of specifically.

N_1 is number of labels with $m = 1$

N_2 is number of labels with $m = 2$

$$\Psi(N_1, N_2, N_3 \ldots) = \sum P\, \psi_{m_i}(x_i)$$

the wave function has an infinite number of labels (eventually they are all zero).

The normalization is

$$\int \left| \sum P\, \psi^2 \right| dx_1 \ldots dx_N = \frac{N!}{N_1!\, N_2!\, N_3! \ldots}$$

since only terms which have the same P will give a nonzero integration, which is one. The easiest way to see this is to write out an example.

Hence the normalized total (generic) wave function is

$$\Psi_{N_1 N_2 N_3 \cdots} = \sqrt{\frac{N_1!\, N_2!\, N_3! \ldots}{N!}} \sum P\psi_{m_i}(x_i) \tag{12-4}$$

A matrix element between the one particle operator

$$\varphi = \sum \varphi(i)$$

is $< \Psi_{N_1 N_2} \cdot \; | \, Q \, | \; \Psi_{N_1 N_2} \ldots >$

$$= Q_{mm'} \sqrt{N_m N_{m'}} < N_1 N_2 \ldots N_m - 1, |Q| N_1 \; \ldots \; N_{m'} - 1 \ldots > \tag{12-5}$$

We define now an operator:

a_m = remove one wave function with label m from the sum of products.

$$a_m \; \psi(N_1 N_2 \; \ldots) = \sqrt{N_m} \;\; \Psi(N_1 \; \ldots \; N_m - 1) . \tag{12-6}$$

(If the operator appears on the right it will be the conjugate a^+. With the help of this operator the matrix element can be written.

$$< \Psi | Q | \psi > = \sum Q_{mm'} < \psi \; | a_m^+ a_{m'} | \psi > \tag{12-7}$$

The matrix by operators

$$Q \rightarrow \sum Q_{mm'} \;\; a_m^+ \; a_{m'} \tag{12-8}$$

If we have a two particle interaction (pair interaction) four wave functions are affected and we have.

$$V(r - r^1) \rightarrow V_{klmn} \, a_k^+ \, a_l^+ \, a_n \, a_m \tag{12-9}$$

Note that the labels n and m are are interchanged. (A convention that is only of importance for fermions.)

The description of fermions, can be given along the same lines. The many particle antisymmetrized wave function is the Slater determinant. After appropriate normalization one can determine the matrix element of a (symmetric) operator.

The operation: ''remove one wave function''

$$a_m \Psi(N_1 \ldots N_m \ldots) = \psi(N_1 \ldots N_m - 1 \ldots) \quad (12\text{-}10)$$

and "insert one wave function":

$$a_m^+ \Psi(N_1 \ldots N_m \ldots) = \psi_1(N_1 \ldots N_m + 1 \ldots) \quad (12\text{-}11)$$

should have a number operator that has only two eigenvalues

$$a_m^+ a_m = 0 \quad \text{or} \quad 1$$

The same relation which exists between the boson operators and operator algebra of the harmonic oscillator exists between the fermion operator and the algebra of the spin = $\frac{1}{2}$ problem.

We associate

$$a^+ = S_x + i\, S_y = \begin{pmatrix} 0 & 1 \\ 0 & 0 \end{pmatrix} \; ; \; a = S_x - i\, S_y = \begin{pmatrix} 0 & 0 \\ 1 & 0 \end{pmatrix}$$

hence

$$a^+ a = \begin{pmatrix} 1 & 0 \\ 0 & 0 \end{pmatrix} \qquad \text{that is } n = 0 \text{ or } 1$$

with:

$$a^+ a + a a^+ = 1 \qquad (a^+)^2 = (a)^2 = 0$$

13 EVALUATION OF THE TIME ORDERED INTEGRAL.

After second quantization we can write the integrand of S_n as a sum of integrals.

Example of $n = 3$:

$$\frac{(-1)^3}{3!} \sum_{\substack{12 \\ \text{summation}}} V_{k_3 l_3 m_3 n_3} \, V_{k_2 l_2 m_2 n_2} \, V_{k_1 l_1 m_1 n_1} \, I \quad (13\text{-}1)$$

where the integral is

$$I = \int \cdots \int a_{k_3}^+(u_3)\, a_{l_3}^+(u_3)\, a_{m_3}(u_3)\, a_{k_2}^+(u_2) \ldots a_{n_1}(u_1) \rho(\beta)\, du_3\, du_2\, du_1 \quad (13\text{-}2)$$

in which all operators are functions of $\{u_n\}$. Originally we assigned the interaction representation exponentials to each set of four operators, but one can assign them individually just as well:

$$e^{u_3\mathcal{H}_0} a_k^+(0)\; a_l^+(0)\; a_m(0)\, a_n(0)\, e^{-u_3\mathcal{H}_0} \tag{13-3}$$

$$= \underbrace{e^{u_3\mathcal{H}_0} a_k^+(0)\, e^{-u_3\mathcal{H}}}\; \underbrace{e^{u_3\mathcal{H}} a_l^+(0)\, e^{-u_3\mathcal{H}}}\; \underbrace{e^{u_3\mathcal{H}} a_m\, e^{-u_3\mathcal{H}}}\; \underbrace{e^{u_3\mathcal{H}} a_n\, e^{-u_3\mathcal{H}}}$$

The individual interaction representation can be simplified and the operator becomes a product of a c-number under the integral and an operator outside the integral: We have

$$a_k^+(u) = e^{u\mathcal{H}_0} a_k^+(0)\, e^{-u\mathcal{H}_0} = e^{u\Sigma\epsilon_l n_l}\, a_k^+(0)\, e^{-u\Sigma\epsilon_{l'} n_{l'}} \;(n_l = a_l^+ a_l)$$

$$= e^{u\epsilon_k a_k^+ a_k}\, a_k^+(0)\, e^{u\epsilon_k a_k^+ a_k} \tag{13-4}$$

since a_k^+ commutes with all a_l^+ and a_l $(k \neq l)$.
We have $a_k^+ (a_k^+ a_k) = (n_k - 1)\, a_k^+$ and by repeated use of this relation:

$$a_k^+ (a_k^+ a_k)^p = (n_k - 1)^p\, a_k^+$$

where p is an arbitrary power, hence

$$a_k^+(0)\; e^{-u\epsilon_k a_k^+ a_k} = e^{u\epsilon_k}\, a_k(0) = a_k(0)\, e^{u\epsilon_k}$$

similarly (13-5)

$$a_k(u) = e^{-u\epsilon_k}\, a_k(0)$$

Hence the integration over all variables u can be performed separately. Each set of four operators will contain an exponential of the type

$$V_{klmn}\, a_k^+ a_l^+ a_n\, a_m \rightarrow e^{u(\epsilon_k + \epsilon_l - \epsilon_m - \epsilon_n)} \tag{13-6}$$

The resulting 4p + 1 operators can be "contracted" by means of the following theorem.

14 WICK's THEOREM (FOR $T \neq 0$).

Suppose we have a set of operators a, b, c, ... f, g from which we want to calculate the trace over the unperturbed density matrix

$$\langle abc \ldots g \rangle = \mathrm{tr}\,\{abcd \ldots g\rho\}. \tag{14-1}$$

we write this product as:

$$\begin{aligned}
abcd \ldots fg = {} & abcd \ldots fg - bacd \ldots fg\,\rho + \\
& bacd \ldots fg - bcad \ldots fg\,\rho + \\
& bcad \ldots fg - bcda \ldots fg\,\rho + \\
& \ldots\ldots\ldots\ldots\ldots\ldots\ldots\ldots \\
& \ldots\ldots\ldots\ldots -bcda \ldots fga\,\rho + \\
& bcdafga\rho. \\
= {} & [ab]\ cd \ldots fg\,\rho \\
& b\,[ac]\ d \ldots fg\,\rho \\
& bc\,[ad] \ldots fg\,\rho \\
& \ldots \\
& bcd \ldots [a, g]\,\rho + bcd \ldots fga\,\rho
\end{aligned}$$

the last interchange $a\rho \rightarrow \rho a$ will create an exponential factor. (Compare Section 13.) The last term is brought to the left and the trace is taken at both sides. Now we see that:

1) Since a cyclic permutation under the trace sign is allowed the two terms on the left-hand side can be combined.

2) Since the commutators are c-numbers we can take them outside the trace symbol hence we get

$$(1 - e^{\pm\beta\epsilon_k}) \langle abcd \ldots g \rangle$$

$$= [ab] \langle cd \ldots fg \rangle + [ac] \langle bd \ldots fg \rangle$$

$$+ \ldots + [ag] \langle bcd \ldots f \rangle. \tag{14-2}$$

dividing both sides through the first factor and introducing the "contractor"

$$\overline{ab} = \frac{[ab]}{1 - e^{\pm\beta\epsilon_k}} \, [\text{because } \rho = \rho(\beta)] \tag{14-3}$$

instead of the commutator, we have

$$\langle abcd \ldots \rangle = \overline{ab}\langle cd \ldots fg \rangle + \overline{ac}\langle bd \ldots fg \rangle \ldots + \overline{ag}\langle bcd \ldots f \rangle \tag{14-4}$$

In the contraction symbol the + sign is to be used if the operator a is a_k^+ and—sign if the operator a is a_k. The second operator is irrelevant, but since the commutator is zero if it has the same superscript as a: b has to be a_k in the first case and a_k^+ in the second case in order to obtain a non-zero contraction, hence ab is either

$$\overline{a_k^+ a_k} = \frac{-1}{1 - e^{\beta\epsilon_k}} = \frac{1}{e^{\beta\epsilon_k} - 1} = \langle n_k \rangle_{\text{B. E.}} \tag{14-5}$$

or

$$\overline{a_k a_k^+} = \frac{1}{1 - e^{-\beta\epsilon_k}} = 1 + \langle n_k \rangle_{\text{B. E.}} \tag{14-6}$$

We can repeat this procedure: for instance, in the first term, select c and write down:

$$\langle cd \ldots g \rangle = \overline{cd}\langle e \ldots g \rangle + \overline{ce}\langle d \ldots g \rangle + \tag{14-7}$$

The general theorem is that we have to select all possible sets of pairs and sum over these. If the particles are Fermions the sum has alternating signs:

$$\langle abc \ldots g \rangle = \overline{ab}\langle c \ldots g \rangle - \overline{ac}\langle b \ldots g \rangle + \overline{ad}\langle abe \ldots g \rangle - \tag{14-8}$$

and the contraction is defined as

$$\overline{ab} = \frac{[ab]}{1 + e^{\pm\beta\epsilon_k}} \tag{14-9}$$

If particles are neither Fermion nor bosons (quasiparticles or spins) the $a^+\rho$ versus ρa^+ relation becomes complicated.

15 DIAGRAMS.

Consider the first order term: $(V_{klmn} \equiv \langle kl|V|mn\rangle)$

$$V_{klmn} \langle a_k^+ a_l^+ a_n a_m \rangle \qquad (15\text{-}1)$$

Wick's theorem gives, upper contraction is the first term, lower contraction the second

$$\langle kl\,|V|kl\rangle \langle a_k^+ a_k\rangle \langle a_l^+ a_l\rangle +$$

$$+ \langle kl\,|V|lk\rangle \langle a_k^+ a_k\rangle \langle a_l^+ a_l\rangle. \qquad (15\text{-}2$$

momentum conservation $\delta(k + l - m - n)$ is taken care of. Each $\langle kl\,|V|mn\rangle$ is represented by a horizontal dotted line. The two ends carry four lines: the two at left are labeled k and m (associated with r) in

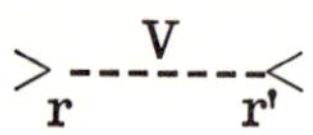

the two at right are labeled l and n (associated with r'). The creation operators a_k^+ and a_l^+ are leaving, the annihilation a_m and a_n coming in:

either: 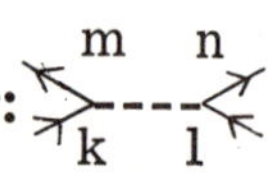

or: 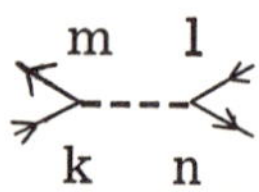

The contraction is indicated by a connection of the two operator lines involved. Hence the lower contraction is:

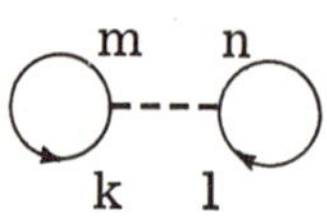

and the upper is
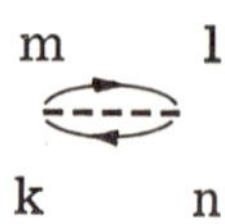

A contraction is a line in the sense that it picks up two loose ends and connects them. The Brueckner-Goldstone diagrams have in common with the Mayer diagrams that a line represents a connection (in the first case a contraction and in the second case an integral). However there is a big difference: the lines in the Mayer diagrams represent the interaction and the circles are reference point, but in the B. G. diagrams the drawn lines are for reference. They are, so to speak, dual diagrams to the first ones. For a more complete comparison of different diagrams, see Morita's lectures.
The next term (second order term) contains 12 possible contractions.

a) four are obtained by repeating the steps of the first order term on each quadruple separately.

b) four entirely new type of contractions

1. 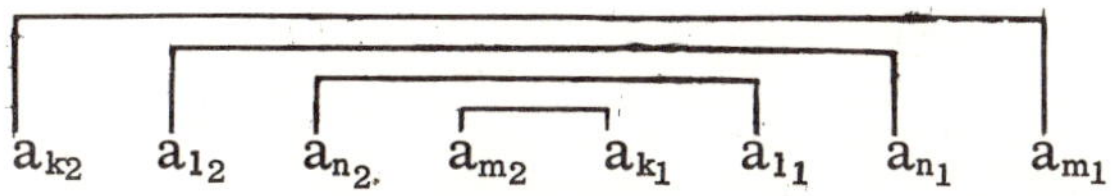

2. 

3. 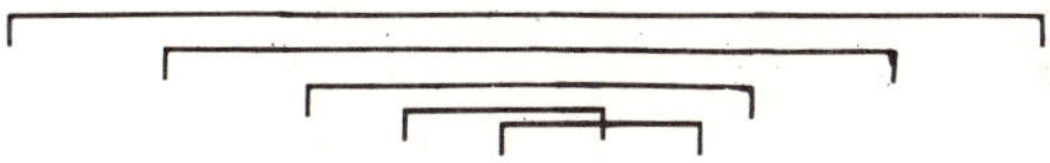

4. 

c) plus four that are a "mixture" of the type a) and the type b). (See diagrams below).

Now we use two horizontal dotted lines, the "later" one is higher. The set a) is represented by

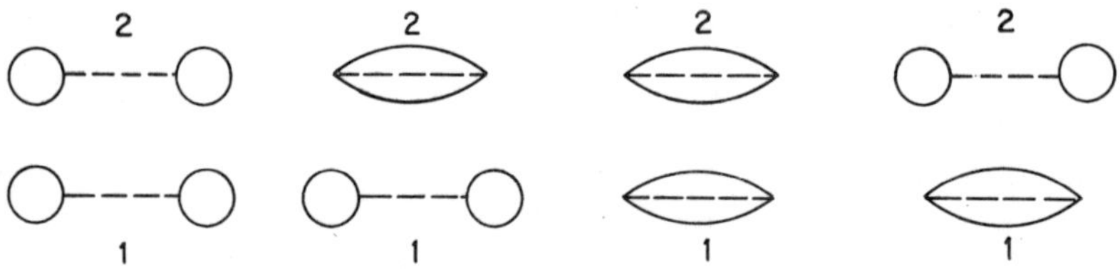

the set b) consists of the following 4:

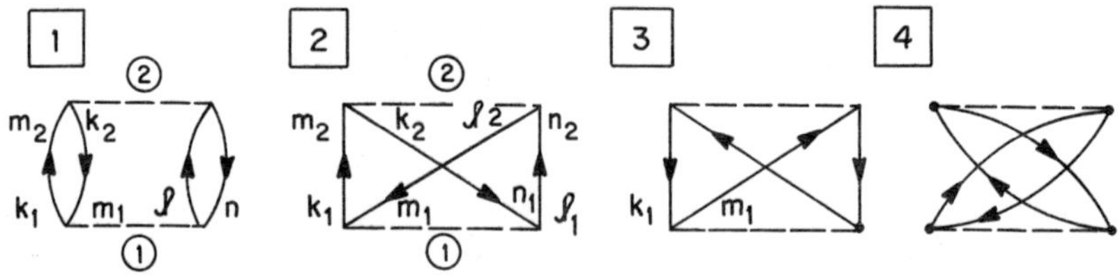

and finally the third set of four:

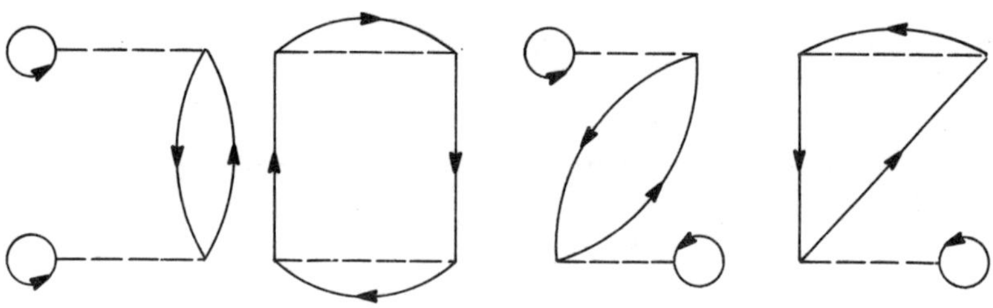

Since an operator with u_2 was always left of one with u_1 all contractions $\overline{a^{+}(u_2)\; a}(u_1)$ are going down (and give a factor $\langle n \rangle$). The ones which have $\overline{aa}^{+}$ are going up (factor: $1 + \langle n \rangle_{BE}$)

Example of one particular term. Take the diagram corresponding to b_1 since $k_2 = m_1$; $l_2 = n_1$; $n_2 = l_1$; $m_2 = k_1$ we will have matrix elements $\langle kl | V | mn \rangle \langle mn | V | lk \rangle$ and the contractions give a factor $1 + \langle n \rangle$ for the first two (lines are going up) and $\langle n \rangle$ for the last two (lines are going down).

We have

$$\sum_{klmn} (1+\langle n_k\rangle)(1+\langle n_l\rangle)\langle n_m\rangle\langle n_n\rangle\langle kl|V|mn\rangle\langle mn|V|lk\rangle I \tag{15-3}$$

where the integral is

$$I_{klmn} = \int_0^\beta du_2 \int_0^{u_2} du_1 \, e^{u_1\Delta_1} e^{u_2\Delta_2} \tag{15-4}$$

one can also "read" the values of Δ_1 from the diagram. Draw a line just above the dotted line number 1.

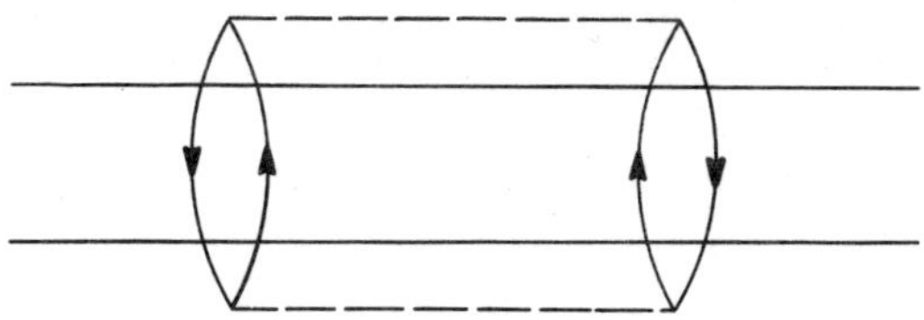

Every line that is leaving "takes away" $+\epsilon$, the ones that are entering carry $-\epsilon$. Hence

$$\Delta_1 = \epsilon_k + \epsilon_l - \epsilon_m - \epsilon_n \, .$$

and

$$\Delta_2 \,(= - \Delta_1 = E) = \epsilon_m + \epsilon_n - \epsilon_k - \epsilon_l$$

$$\int_0^\beta du_2 \int_0^{u_1} du_1 \, e^{-u_2 E} e^{u_1 E} = \int_0^\beta du_2 \, \frac{1 - e^{-u_1 E}}{E}$$

$$= \frac{1}{E^2} (e^{\beta E} - 1 + E\beta). \tag{15-5}$$

Diagram b_4 gives the same contribution since m and n, l and k are interchanged in the second matrix element. It is the same because in the definition of $\langle kl|V|mn\rangle$ it does not matter whether integration over r and r' are interchanged.

It turns out that the diagrams b_2 and b_3 give equal contributions too, but for a different reason: this is a special example of a rule that, the horizontal lines can be cyclically

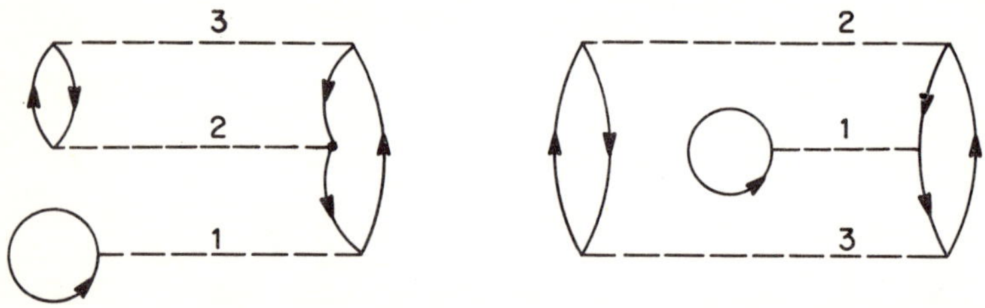

permuted, by changing the variables u_3, u_2 and u_1 into $u_3^1 = u_3 - \beta$, $u_2^1 = u_2$, $u_1^1 = u_1$, gives: $u_2^1 > u_1^1 > u_3^1$ instead of $u_3 > u_2 > u_1$ hence the diagram can be put on a cylinder,

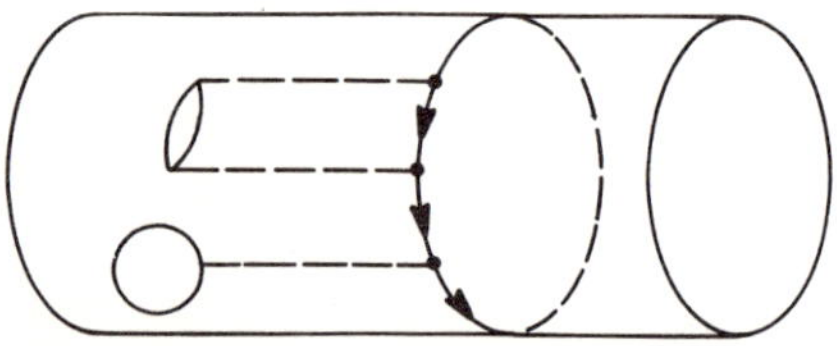

in order to emphasize this equivalence.

16 FINAL REMARKS

1) To calculate the free energy one has to take the logarithm. This process leads to a similar step as the classical Mayer diagrams: the terms consist of connected diagrams only.

2) The formalism for Fermions contains factors $\langle n_k \rangle (1 - \langle n_l \rangle)$, which if k = l will be zero for T = 0. Kohn and Luttinger call the diagrams with this factor (C_3 is an example) anomalous diagrams since they do not occur in field theory. The diagrams explained in these notes are Brueckner-Goldstone diagrams with the exception of these anomalous diagrams.

3) Topics treated with this technique.

a) Self consistent field, Hartree-Fock approximation, which diagrams or sums over diagrams give this type of approximation.
b) He^3 theory and Fermi systems with interaction in general.
c) Superconductivity; although this introduces entirely new elements in the theory.
d) Phonon-phonon interaction in solids. (Van Hove)
e) Plasmas (Montroll & Ward (1958)).
f) Classical density fluctuations (Progogine).
g) Spin-systems (Korringa).

LITERATURE LIST

Page 1 The density matrix was introduced by:

Von Neumann, Göttinger Nachrichten, page 273, 1927.

Dirac, Proc. Camb. Phil. Soc , **25**, 62, 1929.

General references:

Tolman, Chapter IX

Born, Nat. Phil. of Cause and Change, page 193

K. Husimi, Prog. Theor. Phys., **9**, 238, 1953.

Kemble: The Fundamental Principles of Quantum Mechanics.

Fano and Fano, Basic Physics Wiley 1959, N. Y.

Application of Density Matrix not discussed in Notes:

Karplus and Schwinger, Phys. Rev., **73**, 1020, 1948 (line broadening).

Tolhoek and de Groot, Physica, **15**, 833, 1949 (theorem on rotational invariance).

Fano, U. Rev. Mod. Phys., (expansion in complete set of operators.)

Page 7 E. Wigner, Phys. Rev., **40**, 749 (1932).

Irving and Zwanzing, R. J. Chem. Phys., **19**, 1173, 1951.

Page 12 K. Husimi, Proc. Phys. Math. Japan **22**, 264 (1940).

Page 16 Bloch F., Z. f. Physik, **74**, 295, 1932.

Page 17 Husimi: 1940 l. c.

Sommerfeld: Differential Equations. Acad. Press.

Page 19 Dyson, F., Phys., Rev. **75**, 486, 1949.

(General reference; although mainly concerned with Fermions):

D. J. Thouless, The Quantum Mechanics of Many Body Systems. Academic Press.

Page 21 R. Kubo, J. Chem. Phys., **20**, 770, 1952.

Gell-mann, Goldberger, Phys. Rev., **91**, 389, 1953.

Page 26 L. D. Landau and E. M. Lifshitz, Quantum Mechanics (Vol. 3, Theoretical Physics) Sec. 62.

Page 28 Condon and Shortley. Theory of atomic spectra page 171.

Page 32 Wick's theorem for T = 0:

Thouless, M. Phys. Rev., **107**, 1162, 1957.

Bloch, C. and Dominicis, Nuclear Physics **7**, 459, 1958 and **10**, 181, 1959.

Final version: Gaudin, M. Nuclear Physics, **15**, 89

See also: C. de Dominicis: Methodes Diagrammatiques en Machanique Statistique Quantique C. E. A No. 1873.

Page 38 A number of crucial papers are collected in Intern. Rev. Series No. 2 by H. Morrison-Gordon and Breach 1963. Also: L. van Hove, Hugenholtz and Howland-W. A Benjamin., Inc. Both are called: Quantum Theory of Many Particle System.

Method of Green's Function

Tomoyasu Tanaka

1.1 BASIC DEFINITIONS

$\psi^+(r)$: The creation operator, when acting to the right on a state of a system, adds a particle to the state at the space-time point r, t.

$\psi(r)$: The annihilation operator, the adjoint of the creation operator, acting to the right, removes a particle from the state at the point r, t.

The density of particles at the point* r, t is

$$n(r, t) = \psi^+(r, t)\, \psi(r, t).$$

The total number of particles in the system is given by

$$N(t) = \int dr\, \psi^+(r, t)\, \psi(r, t)$$

The total energy of a system of particles of mass m interacting through an instantaneous two-body potential v(r) is given by

$$H(t) = \int dr \frac{\nabla\psi^+(r,t) \cdot \nabla\psi(r,t)}{2m} \qquad [\hbar = 1]$$
$$+ \frac{1}{2}\int dr\, dr'\, \psi^+(r,t)\, \psi^+(r',t)\, v(|r - r'|)\, \psi(r',t)\, \psi(r,t) \tag{1-1}$$

---

* The symbol r stands nearly everywhere for the three dimensional vector r.

The equation of motion [Heisenberg picture]

$$i\frac{\partial X(t)}{\partial t} = [X(t),\ H(t)] \rightarrow X(t) = e^{iHt} X(0) e^{-iHt} \qquad (1\text{-}2)$$

$$[H(t),\ H(t)] = 0, \qquad [N(t),\ H(t)] = 0.$$

The commutation relations

$$\psi(r, t)\ \psi(r', t) \mp \psi(r', t)\ \psi(r, t) = 0 \qquad \text{-B. E.}$$

$$\psi^{+}(r, t)\,\psi^{+}(r', t) \mp \psi^{+}(r', t)\,\psi^{+}(r, t) = 0 \qquad \text{+F. D.} \qquad (1\text{-}3)$$

$$\psi(r, t)\ \psi^{+}(r', t) \mp \psi^{+}(r', t)\ \psi(r, t) = \delta(r - r')$$

The average of an operator over the grand canonical ensemble

$$<X> = \frac{\mathrm{tr}\,[e^{-\beta(H-\mu N)}X]}{\mathrm{tr}[e^{-\beta(H-\mu N)}]} \quad \mathrm{tr}[e^{-\beta(H-\mu N)}\,X]$$

$$= \sum_{N=0}^{\infty} e^{+\mu N\beta} \sum_{i} <i|X|i>_N \ e^{-\beta E_i(N)} \qquad (1\text{-}4)$$

THE GREEN'S FUNCTIONS

The one-particle Green's function is defined by

$$G(1.\,1') = -i < T\psi(1)\,\psi^{+}(1') > \qquad (1\text{-}5)$$

The two-particle Green's function

$$G_2(1\,2,\ 1'\,2') = i^2 < T\psi(1)\,\psi(2)\,\psi^{+}(2')\,\psi^{+}(1') > \qquad (1\text{-}6)$$

where

$$\psi(1) = \psi(r_1,\ t_1)\ .\quad \psi^{+}(1') = \psi^{+}(r'_1,\ t'_1),$$

and T represents the time ordering operator.
When T is applied to a product of operators, it arranges

them in chronological order with the earliest time appearing on the right and the latest on the left. For bosons, this is the full effect of T. For fermions, however, it is convenient to define T to include an extra factor ± 1, depending on whether the resulting time-ordered product is an even or odd permutation of the original order.

For example,

$$T(\psi(1)\psi^+(1')) = \psi(1)\psi^+(1') \quad \text{for } t_1 < t'_1.$$
$$= \pm\psi^+(1')\psi(1) \quad \text{for } t_1 < t'_1.$$

The one-particle Green's function has a direct physical interpretation. It describes the propagation of disturbances in which a single particle is either added to or removed from the many-particle equilibrium system. For example, when $t_1 > t'_1$, the creation operator acts first, producing a disturbance by adding a particle at the space-time point r'_1, t'_1. This disturbance then propagates to the later time t_1, when a particle is removed at r_1 ending the disturbance and returning the system to its equilibrium state.

PROOF:

$$T\psi(1)\psi^+(1') = \theta(t_1 - t'_1)\psi(1)\psi^+(1') \pm \theta(t'_1 - t_1)\psi^+(1')\psi(1)$$

$$i\frac{d}{dt_1}[T\psi(1)\psi^+(1')] = i\left[\frac{d}{dt_1}\theta(t_1 - t'_1)\right]\psi(1)\psi^+(1')$$
$$\pm i\left[\frac{d}{dt_1}\theta(t'_1 - t_1)\right]\psi^+(1')\psi(1)$$
$$+\theta(t_1 - t'_1)\, i\frac{d\psi(1)}{dt_1}\psi^+(1') \qquad (1\text{-}7)$$
$$\pm\theta(t'_1 - t_1)\psi^+(1')\, i\frac{d\psi(1)}{dt}$$

$$= i\,\delta(t_1 - t'_1)[\psi(1)\psi^+(1') \mp \psi^+(1')\psi(1)]$$

$$- H^x(1)[\theta(t_1 - t'_1)\psi(1)\psi^+(1') \pm \theta(t'_1 - t_1)\psi^+(1')\psi(1)]$$

(where $H^x(1)\psi(1) = H(1)\psi(1) - \psi(1)H(1) = [H(1), \psi(1)]$)

$$= i\,\delta(t_1 - t'_1)\,\delta(r_1 - r'_1) - H^x(1)\,T\psi(1)\psi^+(1')$$

Multiplying $-i$
and $<>$

$$\left[i\frac{d}{dt_1} + H^x(1)\right] G(1.\,1') = \delta(t_1 - t'_1)\,\delta(r_1 - r'_1) \qquad (1\text{-}8)$$

end of proof.

Similarly, the two-particle Green's function describes, for the various time orders disturbances produced by the removal or addition of two particles.

In addition to the one-particle Green's function we define the correlation functions

$$(1\text{-}9)$$

$$G^>(1.\,1') = -i\langle\psi(1)\psi^+(1')\rangle,\quad G(1.1') = G^>(1.\,1'),\quad t_1 > t'_1,$$

$$G^<(1.\,1') = \pm(1/i)\langle\psi^+(1')\psi(1)\rangle,\quad G(1.1') = G^<(1.1'),\quad t_1 < t'_1.$$

1. 2 THE BOUNDARY CONDITION

$$G^>(1.\,1') = \frac{tr[\psi(1)\psi^+(1')\,e^{-\beta(H-\mu N)}]}{i\,tr[e^{-\beta(H-\mu N)}]}$$

$$= \frac{tr[e^{-\beta(H-\mu N)}e^{iHt_1}\psi(r_1, 0)\,e^{-i(t_1 - t'_1)H}\psi^+(r'_1, 0)\,e^{-it'_1 H}]}{i\,tr[e^{-\beta(H-\mu N)}]}$$

$$(1\text{-}10)$$

$G^>(1.\,1')$ is an analytic function for complex values of the time arguments in the region $0 > \mathrm{Im}(t_1 - t'_1) > -\beta$.

PROOF:

$$e^{-\beta(H-\mu N)}\, e^{i(t_1-t_1')H}\, \psi(r_1,\ 0)\, e^{-i(t_1-t_1')H}\, \psi^+(r_1',0)$$

$$= e^{\beta\mu N}\, e^{-H(\beta-i[t_1-t_1'])}\, \psi(r_1,\ 0)\ \ e^{-i(t_1-t_1')H}\, \psi^+(r_1',0)$$

$E_i(N) \to \infty$ no upper bound

$$\mathrm{Re}(\beta - i[t_1 - t_1']) > 0, \to \beta + \mathrm{Im}(t_1 - t_1') > 0$$

$$\mathrm{Re}[i(t_1 - t_1')] > 0, \qquad \mathrm{Im}(t_1 - t_1') < 0$$

$$\therefore -\beta < \mathrm{Im}(t_1 - t_1') < 0$$

end of proof.

The relation between $G^>$ and $G^<$.

$$G^<(1,1')\Big|_{t_1=0} = \pm\frac{1}{i}\,\frac{\mathrm{tr}[e^{-\beta(H-\mu N)}\,\psi^+(r_1',\ t_1')\ \psi(r_1,\ 0)]}{\mathrm{tr}[e^{-\beta(H-\mu N)}]}$$

$$= \pm\frac{1}{i}\,\frac{\mathrm{tr}[\psi(r_1,0)\,e^{-\beta(H-\mu N)}\,\psi^+(r_1',\ t_1')]}{\mathrm{tr}[e^{-\beta(H-\mu N)}]}$$

$$= \pm\frac{1}{i}\,\frac{\mathrm{tr}[e^{-\beta(H-\mu N)}\{e^{+\beta(H-\mu N)}\,\psi(r_1,0)\,e^{-\beta(H-\mu N)}\,\psi^+(r_1',t_1')\}]}{\mathrm{tr}[e^{-\beta(H-\mu N)}]}$$

$$= \pm\frac{1}{i} < e^{\beta(H-\mu N)}\quad \psi(r_1,\ 0)\ \ e^{-\beta(H-\mu N)}\,\psi^+(r_1',\ t_1') >$$

Now

$$\psi(r_1,\ 0)\ N = \psi(r_1,0)\int dr\,\psi^+(r,0)\,\psi(r,0)$$

$$= \int dr\,\psi(r_1,0)\,\psi^+(r,0)\,\psi(r,\ 0)$$

$$= \int dr[\pm\,\psi^+(r,\ 0)\,\psi(r_1,\ 0) + \delta(r - r_1)]\,\psi(r,\ 0)$$

$$= \int dr[\psi^+(r,\ 0)\,\psi(r,\ 0) + \delta(r - r_1)]\,\psi(r_1,\ 0)$$

$$= (N+1)\ \psi(r_1,\ 0)$$

$$\psi(r_1, 0)\ f(N) = f(N+1)\ \psi(r_1, 0)$$

$$\psi(r_1, 0)\ e^{-\beta(H-\mu N)} = e^{\beta\mu(N+1)}\ \psi(r_1, 0)\ e^{-\beta H}$$

$$G^{<}(1, 1')\Big|_{t_1=0} = \pm\frac{1}{i}< e^{\beta\mu}\ e^{\beta H}\ \psi(r_1, 0)\ e^{-\beta H}\ \psi^{+}(r_1', t_1') >$$

$$e^{\beta H} = e^{i(-i\beta)H}$$

$$= \pm\frac{1}{i}< e^{\beta\mu}\ \psi(r_1, -i\beta)\,\psi^{+}(r_1', t_1') >$$

$$G^{<}(1, 1')\Big|_{t_1=0} = \pm\, e^{\beta\mu}\ G^{>}(1, 1')\Big|_{t_1=-i\beta} \tag{1-11}$$

Now it is possible to define the Green's function in the complex time domain, if we extend the definition of the time-ordering symbol T to mean " $i\times t$" ordering when the times are imaginary. The further down the imaginary axis a time is, the "later" it is Then the Green's function is well defined in the interval $0 \le it \le \beta$.

$$\begin{aligned} G(1, 1') &= < T\psi(r_1, it_1)\ \psi^{+}(r'_1, it'_1)> \\ &= G^{>}(1, 1'), \quad \text{for } it_1 > it'_1, \\ &= G^{<}(1, 1'), \quad \text{for } it_1 < it'_1. \end{aligned} \tag{1-12}$$

For $0 < it'_1 < \beta$, we have

$$G(1, 1')\Big|_{t_1=0} = G^{<}(1, 1')\Big|_{t_1=0} \quad (\text{since} 0 = it_1 < it'_1 \text{ for all } t'_1)$$

and

$$G(1, 1')\Big|_{t_1=-i\beta} = G^{>}(1, 1')\Big|_{t_1=-i\beta} (\text{since } \beta = it_1 > it_1 \text{ for all } t'_1)$$

Therefore the previous result can be restated as a relation between the values of G(1, 1') at the boundaries of the imaginary time domain:

$$G(1, 1')\Big|_{t_1=0} = \pm e^{\beta\mu} G(1, 1')\Big|_{t_1=-i\beta} \tag{1-13}$$

Similarly

$$G_2(12, 1'2')\Big|_{t_1=0} = \pm e^{\beta\mu} G_2(12, 1'2')\Big|_{t_1=-i\beta}$$

$$G_2(12, 1'2')\Big|_{t_1'=0} = \pm e^{\beta\mu} G_2(12, 1'2')\Big|_{t_1'=-i\beta}$$

Because of the translational and rotational invariance of the Hamiltonian in space and its translational invariance in time, $G^>$ and $G^<$ depend only on $|r_1 - r'_1|$ and $t_1 - t_1'$.

$$G(1, 1') = G(|r_1 - r_1'|, t_1 - t_1')$$

In terms of the difference variables, the boundary condition can be written as

$$G^<(r, t) = \pm e^{\beta\mu} G^>(r, t - i\beta). \tag{1-14}$$

We now introduce the Fourier transforms of $G^>$ and $G^<$, defined by (Note the introduction of a factor i in the Fourier transform).

$$G^>(p, \omega) = i \int dr \int_{-\infty}^{\infty} dt\, e^{-ip\cdot r + i\omega t}\, G^>(r, t) \tag{1-15}$$

$$G^<(p, \omega) = \pm i \int dr \int_{-\infty}^{\infty} dt\, c^{-ip\cdot r + i\omega t}\, G^<(r, t) \tag{1-16}$$

Introducing (1-14) into (1-16)

$$G^<(p, \omega) = ie^{\beta\mu} \int dr \int_{-\infty}^{\infty} dt\, e^{-ip\cdot r + i\omega t}\, G^>(r, t - i\beta)$$

$$t - i\beta = t'$$

$$= e^{-\beta(\omega-\mu)} G^>(p, \omega). \tag{1-17}$$

It is useful to introduce the "spectral function" $A(p, \omega)$ defined by

$$A(p,\ \omega) = G^{>}(p,\ \omega) \mp G^{<}(p,\ \omega) \tag{1-18}$$

$$= G^{>}(p,\ \omega)\ [1 \mp e^{-\beta(\omega-\mu)}]$$

$$= G^{<}(p,\ \omega)\,[e^{\beta(\omega-\mu)} \mp 1]$$

$$G^{<}(p,\ \omega) = f(\omega)\ A(p,\ \omega) \tag{1-19}$$

where

$$f(\omega) = \frac{1}{e^{\beta(\omega-\mu)} \mp 1} \tag{1-20}$$

$$G^{>}(p,\ \omega) = \frac{1}{1 \mp e^{-\beta(\omega-\mu)}}\ A(p,\ \omega)$$

$$= e^{\beta(\omega-\mu)}\, f(\omega)\, A(p,\ \omega)$$

$$= [\ 1 \pm f(\omega)\ ]\, A(p,\ \omega) \tag{1-21}$$

From (1-18), (1-19) and (1-20)

$$A(p,\ \omega) = i \int dr \int_{-\infty}^{\infty} dt\ e^{-ip\cdot r + i\omega t}\ [G^{>}(r,\ t) - G^{<}(r,\ t)]$$

$$= \int dr \int_{-\infty}^{\infty} dt\ e^{-ip\cdot r + i\omega t} < \psi(r,\ t)\, \psi^{+}(0,\ 0) \mp \psi^{+}(0,\ 0)\, \psi(r,\ t) >$$

$$\delta(t) = \frac{1}{2\pi} \int_{-\infty}^{\infty} e^{i\omega t}\ d\omega$$

$$\int \frac{A(p,\omega)}{2\pi}\ d\omega = \int dr\ e^{-ip\cdot r} < [\psi(r,0)\, \psi^{+}(0,0) \mp \psi^{+}(0,0)\, \psi(r,0)] >$$

$$= \int dr\ \delta(r) = 1. \qquad \text{(sum rule)} \tag{1-22}$$

EXAMPLE

Ideal gas

$$H_0 = \int dr \, \frac{\nabla\psi^+(r,t)\cdot\nabla\psi(r,t)}{2m}$$

$$= \int dp \, \frac{p^2}{2m} \, \psi^+(p,t) \; \psi(p,t) \qquad (1\text{-}23)$$

$$G^<(p, \omega) = (2\pi)^3 \int dt \frac{e^{i\omega t}}{\Omega} < \psi^+(p, 0)\,\psi(p, t) > \qquad (1\text{-}24)$$

$$\psi(p, t) = e^{iHt}\,\psi(p, 0)\,e^{-iHt} = e^{-i(p^2/2m)t}\,\psi(p, 0) \qquad (1\text{-}25)$$

$$G^<(p, \omega) = (2\pi/\Omega)\;\delta(\omega - \frac{p^2}{2m})<\psi^+(p, 0)\,\psi(p, 0)> \qquad (1\text{-}26)$$

From (1-19)

$$A(p, \omega) \propto G^<(p, \omega) \sim \delta(\omega - \frac{p^2}{2m})$$

$A(p, \omega) = K\;\delta(\omega - \frac{p^2}{2m})$, Then from (1-23)

$$A(p, \omega) = 2\pi\delta(\omega - \frac{p^2}{2m}) \qquad (1\text{-}27)$$

From (1-15) and (1-16)

$$G^>(r, t) = \frac{1}{i}\cdot\frac{1}{(2\pi)^4}\int dp \int d\omega \; G^>(p, \omega)\,e^{ip\cdot r - i\omega t}$$

(1-21), (1-27)

$$= \frac{1}{i}\,\frac{1}{(2\pi)^4}\int dp \int d\omega \; e^{ip\cdot r - i\omega t}\,[1 \pm f(\omega)\,]\,(2\pi)^4\,\delta(\omega - \frac{p^2}{2m})$$

$$= \frac{1}{i}\int dp \; e^{ip\cdot r - i(p^2/2m)t}[1 \pm f(p^2/2m)\,] \qquad (1\text{-}28)$$

From (1-14)

$$G^{<}(r, t) = \pm e^{\beta\mu} G^{>}(r, t - i\beta)$$

$$= \pm\frac{1}{i} \int dp\, e^{ip\cdot r - i(p^2/2m)t} e^{\beta(\mu - \frac{p^2}{2m})}[1 \pm f(p^2/2m)]$$

$$= \pm\frac{1}{i} \int dp e^{ip\cdot r - i(p^2/2m)t} f(p^2/2m) \tag{1-29}$$

From (1-19), (1-26) and (1-27)

$$\langle n(p) \rangle = \frac{\langle \psi^{+}(p, 0)\psi(p, 0)\rangle}{\Omega} = f(p^2/2m)$$

$$= \frac{1}{e^{\beta(p^2/2m - \mu)} \mp 1} \tag{1-30}$$

Derivation of (1-23) ~(1-26)

The commutation relations:

$$\psi(p, t) = \frac{1}{(2\pi)^{3/2}} \int dr\, e^{-ip\cdot r} \psi(r, t),$$

$$\psi^{+}(p, t) = \left(\frac{1}{2\pi}\right)^{3/2} \int dr\, e^{ip\cdot r}\, \psi^{+}(r, t)$$

$$\psi(p, t)\psi^{+}(p', t) = \frac{1}{(2\pi)^3} \int dr \int dr'\, e^{-ip\cdot r + ip'\cdot r'} \psi(r, t)\psi^{+}(r', t)$$

$$= \frac{1}{(2\pi)^3} \int dr \int dr'\, e^{-ip\cdot r + ip'\cdot r'} [\pm\psi^{+}(r', t)\psi(r, t) + \delta(r - r')]$$

$$= \pm\psi^{+}(p', t)\psi(p, t) + (2\pi)^3\, \delta(p - p')$$

$$\psi(p, t)\psi^{+}(p', t) \mp \psi^{+}(p', t)\psi(p, t) = \delta(p - p')$$

Similarly

$$\psi^+(p,t)\,\psi^+(p',t) \mp \psi^+(p',t)\;\psi^+(p,t) = 0$$

$$\psi(p,t)\,\psi(p',t) \mp \psi(p',t)\;\psi(p,t) = 0$$

$$H_0 = \int dr\,\frac{\nabla\psi^+(r,t)\cdot\nabla\psi(r,t)}{2m}$$

$$= \frac{1}{(2\pi)^3}\int dp\int dp'\,\frac{(p\cdot p')}{2m}\int e^{i(p-p')\cdot r}\psi^+(p,t)\,\psi(p',t)\,dr$$

$$= \int dp\,\frac{p^2}{2m}\;\psi^+(p,t)\;\psi(p,t)$$

From (1-30)

$$G^<(p,\omega) = \pm i\int dr\int dt\;e^{-ip\cdot r+i\omega t}\;G^<(r,t)$$

$$= \int dr\int dt\;e^{-ip\cdot r+i\omega t}\;\langle\psi^+(0,0)\,\psi(r,t)\rangle$$

$$= \frac{1}{\Omega}\int dr_2\int dr_1\int dt\;e^{-ip\cdot(r_1-r_2)+i\omega t}$$

$$\langle\psi^+(r_2,0)\,\psi(r_1,t)\rangle$$

$$= \frac{(2\pi)^3}{\Omega}\int dt\;e^{i\omega t}\langle\psi^+(p,0)\,\psi(p,t)\rangle$$

$$\psi(p,t) = e^{iHt}\,\psi(p,0)\,e^{-iHt}$$

$$\psi(p,0)\;H = \psi(p,0)\;\int dp_1\,\frac{p_1^2}{2m}\;\psi^+(p,\theta)\,\psi(p,\theta)$$

$$= \int dp_1\,\frac{p_1^2}{2m}\left[\pm\psi^+(p_1,\theta)\,\psi(p,\theta) + \;\delta(p_1-p)\,\right]\psi(p_1,\theta)$$

$$= \int dp_1 \frac{p_1^2}{2m} [\psi^+(p_1, \theta)\psi(p_1, \theta) + \delta(p_1 - p)]\psi(p, \theta)$$

$$= (H + \frac{p_1^2}{2m}) \psi(p, 0)$$

$$\psi(p, 0) f(H) = f(H + \frac{p^2}{2m}) \psi(p, 0)$$

$$\psi(p, 0) e^{-iHt} = e^{-i(H+\frac{p^2}{2m})} \psi(p, 0)$$

$$\therefore \quad G^<(p, \omega) = \frac{(2\pi)^3}{\Omega} \int dt\, e^{i(\omega - \frac{p^2}{2m})t} \langle \psi^+(p, 0)\psi(p, 0)\rangle$$

$$= \frac{(2\pi)^4}{\Omega} \delta(\omega - \frac{p^2}{2m}) \langle \psi^+(p, 0)\psi(p, 0)\rangle$$

1.3 STATISTICAL MECHANICAL INFORMATION CONTAINED IN G.

The equation of motion for $\psi^+(r, t)$ is

$$i \frac{\partial}{\partial t} \psi(r, t) = [\psi(r, t), H(t)]$$

where

$$H(t) = \int d\bar{r} \frac{\nabla\psi^+(\bar{r}, t) \cdot \nabla\psi(\bar{r} t)}{2m}$$

$$+ \frac{1}{2} \iint dr_1\, dr_2\, V(r_1 - r_2) \psi^+(r_1 t) \psi^+(r_2 t) \psi(r_2 t) \psi(r_1 t)$$

$$[\psi(r t), H(t)]$$

$$= - \int d\bar{r} \frac{1}{2m} [\psi(r t)\psi^+(\bar{r} t) \bar{\nabla}^2 \psi(\bar{r} t) - \psi^+(\bar{r} t)\bar{\nabla}^2 \psi(\bar{r} t)\psi(r t)]$$

$$+ \frac{1}{2} \iint dr_1\, dr_2 [\psi(r t)\psi^+(r_1 t)\psi^+(r_2 t)\psi(r_2 t)\psi(r_1 t)$$

$$- \psi^+(r_1 t)\psi^+(r_2 t)\psi(r_2 t)\psi(r_1 t)\psi(r t)] v(r_1 - r_2)$$

$$\underline{\psi(rt)\psi^+(r,t)}\ \psi^+(r_2 t)\psi(r_2 t)\ \psi(r_1 t)$$

$$= [\pm\psi^+(r_1 t)\underline{\psi(rt)} + \delta(r - r_1)]\ \underline{\psi^+(r_2 t)}\ \psi(r_2 t)\psi(r_1 t)$$

$$= \delta(r - r_1)\psi^+(r_2 t)\ \psi(r_2 t)\psi(rt)$$

$$\pm\psi^+(r_1 t)[\pm\psi^+(r_2 t)\psi(rt) + \delta(r - r_2)]\ \psi(r_2 t)\psi(r_1 t)$$

$$= \delta(r - r_1)\psi^+(r_2 t)\ \psi(r_2 t)\psi(rt)$$

$$\pm\delta(r - r_2)\psi^+(r_1 t)\ \psi(rt)\psi(r_1 t)$$

$$+\psi^+(r_1 t)\ \psi^+(r_2 t)\ \psi(rt)\ \psi(r_2 t)\ \psi(r_1 t)$$

$$= \delta(r - r_1)\ \psi^+(r_2 t)\ \psi(r_2 t)\ \psi(rt)$$

$$+\delta(r - r_2)\psi^+(r_1 t)\psi(r_1 t)\ \psi(rt)$$

$$+\psi^+(r_1 t)\ \psi^+(r_2 t)\ \psi(r_2 t)\ \psi(r_1 t)\ \psi(rt)$$

$$\therefore [\psi(rt),\ H(t)] = -\frac{\nabla^2}{2m}\psi(rt) + \int d\bar{r}\ \psi^+(\bar{r}t)\,v(\bar{r}-r)\,\psi(\bar{r}t)\,\psi(rt)$$

$$\left[i\frac{\partial}{\partial t} + \frac{\nabla^2}{2m}\right]\psi(r,t) = \int d\bar{r}\ \psi^+(\bar{r}t)\,v(\bar{r}-r)\ \psi(\bar{r}t)\ \psi(rt) \tag{1-31}$$

Similarly

$$\left[-i\frac{\partial}{\partial t} + \frac{\nabla'^2}{2m}\right]\psi^+(r',t') = \psi^+(r',t')\int d\bar{r}'\psi^+(\bar{r}',t')\,v(\bar{r}'-r')\,\psi(\bar{r}',t'), \tag{1-32}$$

Therefore it follows that

$$\frac{1}{4}\int dr\left[\left(i\frac{\partial}{\partial t} - i\frac{\partial}{\partial t'}\right)\ \psi^+(r,t')\psi(r,t)\right]_{t'=t}$$

$$= \frac{1}{4}\int dr\left[\left(-\frac{\nabla^2}{2m} - \frac{\nabla'^2}{2m}\right)\psi^+(r',t)\psi(r,t)\right]_{r'=r} \qquad (\nabla' = -\nabla)$$

$$+\frac{1}{2}\int dr\,d\bar{r}\ \psi^+(r,t)\psi^+(\bar{r},t)\,v(r-\bar{r})\,\psi(\bar{r},t)\ \psi(r,t) \tag{1-33}$$

The right side of (1-33) is half the kinetic energy plus all the potential energy.

$$<H> = \frac{1}{4} \int dr \left[\left(i\frac{\partial}{\partial t} - i\frac{\partial}{\partial t'} + \frac{\nabla \cdot \nabla'}{m} \right] \right.$$

$$\left. <\psi^{+}(r', t')\,\psi(r, t)> \right]_{r'=r,\ t'=t}$$

$$= \pm\frac{i}{4} \int dr \left[\left(i\frac{\partial}{\partial t} - i\frac{\partial}{\partial t'} + \frac{\nabla \cdot \nabla'}{m} \right) \right] G^{<}(rt, r'\,t') \Big]_{r'=r,\ t'=t}$$

From (1-16)

$$G^{<}(p, \omega) = \pm i \int dr \int_{-\infty}^{\infty} dt\ e^{-ip.\,r + i\omega t}\ G^{<}(r, t)$$

$$\pm i\, G^{<}(rt, r'\,t') = \frac{1}{(2\pi)^4} \iint dp\, d\omega\ e^{ip\cdot(r-r') - i\omega(t-t')}\ G^{<}(p, \omega)$$

$$\left[\left(i\frac{\partial}{\partial t} - i\frac{\partial}{\partial t'} + \frac{\nabla \cdot \nabla'}{m} \right) e^{ip\cdot(r-r') - i\omega(t-t')} \right]_{r'=r,\ t'=t}$$

$$= \left(2\omega + \frac{p^2}{m}\right)$$

$$<H> = \frac{1}{4} \frac{\Omega}{(2\pi)^4} \iint dp\, d\omega \left(2\omega + \frac{p^2}{m}\right) G^{<}(p, \omega)$$

$$= \frac{\Omega}{(2\pi)^4} \iint dp\, d\omega \left[\frac{\omega + \dfrac{p^2}{2m}}{2} \right] f(\omega)\ A(p, \omega) \qquad (1\text{-}34)$$

This result is useful for evaluating ground-state energy, specific heat, etc.

Relation between the grand partition function and the spectral function.

$$Z_g = \mathrm{tr}\left[e^{-\beta(H-\mu N)} \right] = e^{\beta P \Omega} \qquad (1\text{-}35)$$

For fixed temperature (β) and volume (Ω), we differentiate Z_g with respect to μ.

$$\beta\Omega \frac{\partial P}{\partial \mu}\bigg|_{\beta\Omega} = \frac{\partial}{\partial \mu} \ln Z_g = \frac{\partial}{\partial \mu} \ln \mathrm{tr}\left[e^{-\beta(H-\mu N)}\right]$$

$$= \beta \frac{\mathrm{tr}\left[e^{-\beta(H-\mu N)} N\right]}{\mathrm{tr}\left[e^{-\beta(H-\mu N)}\right]}$$

$$= \beta \langle N \rangle$$

The density of particles is given by

$$N = \frac{\langle N \rangle}{\Omega} = \frac{\partial P}{\partial \mu}\bigg|_{\beta\Omega} \tag{1-36}$$

$\mu \to -\infty, \quad n \to 0\,, \qquad P \to 0.$ (low density limit)

$$P(\beta,\ \mu) = \int_{-\infty}^{\mu} d\mu'\ n(\beta,\ \mu') \tag{1-37}$$

$$n(\beta,\ \mu) = \langle \psi^{+}(r,\ t)\,\psi(r,\ t)\rangle = \pm i\, G^{<}(rt,\ rt)$$

$$= \int \frac{d\omega}{2\pi} \cdot \frac{dp}{(2\pi)^3}\, G^{<}(p,\ \omega)$$

$$= \int \frac{d\omega}{2\pi} \frac{dp}{(2\pi)^3}\ f(\omega)\, A(p,\ \omega)$$

If we know the Green's function as a function of μ, we can calculate P and hence the partition function.
For a free gas

$$A = 2\pi\ \delta\left(\omega - \frac{p^2}{2m}\right)$$

hence

$$n(\beta,\ \mu) = \int \frac{dp}{(2\pi)^3} \frac{1}{e^{\beta\left[\frac{p^2}{2m} - \mu\right]} \mp 1} \tag{1-38}$$

$$P(\beta,\ \mu) = \mp\frac{1}{\beta} \int \frac{dp}{(2\pi)^3} \ln\left\{1 \mp e^{-\beta\left[(p^2/2m) - \mu\right]}\right\} \tag{1-39}$$

In the classical limit $\beta\,\mu \rightarrow -\infty$

$$n = \int \frac{dp}{(2\pi)^3}\, e^{-\beta[\frac{p^2}{2m} - \mu]}$$

$$P = \beta^{-1} \int \frac{dp}{(2\pi)^3}\, e^{-\beta[\frac{p^2}{2m} - \mu]} = \frac{n}{\beta} = \frac{NkT}{\Omega} \;[\text{ideal gas}]$$

The method of coupling constant.

$$H = H_0 + \lambda V. \qquad \lambda \text{ the coupling constant}$$

$$V = \frac{1}{2} \int dr\, d\bar{r}\, \psi^+(r)\, \psi^+(\bar{r})\, v(|r - \bar{r}|)\, \psi(\bar{r})\, \psi(r)$$

$$\frac{\partial}{\partial\lambda} \ln Z_g = \frac{1}{Z_g}\, \mathrm{tr}\left[\frac{\partial}{\partial\lambda} e^{-\beta(H_0 + \lambda V - \mu N)}\right] = -\beta \langle V \rangle_\lambda \qquad (1\text{-}40)$$

Integrating both sides with respect to λ, from $\lambda = 0$ to $\lambda = 1$, we find

$$[\ln Z\,]_{\lambda=1} - [\ln Z\,]_{\lambda=0} = -\beta \int_0^1 \frac{d\lambda}{\lambda} \langle \lambda V \rangle_\lambda \qquad (1\text{-}41)$$

From (1-33) by subtracting half the kinetic energy

$$\langle \lambda V \rangle_\lambda = \Omega \int \frac{dp}{(2\pi)^3} \cdot \frac{d\omega}{2\pi} \cdot \frac{\omega - (p^2/2m)}{2} A_\lambda(p, \omega)\, f(\omega) \qquad (1\text{-}42)$$

So that

$$\beta P \Omega = [\ln Z_g\,]_{\lambda=1}$$

$$= [\ln Z_g\,]_{\lambda=0} - \beta\Omega \int_0^1 \frac{d\lambda}{\lambda} \int \frac{dp}{(2\pi)^3} \cdot \frac{d\omega}{2\pi}\, \frac{\omega - (p^2/2m)}{2} A_\lambda(p, \omega)\, f(\omega) \qquad (1\text{-}43)$$

1. 4. EQUATIONS OF MOTION

(12-1) $\rightarrow$ one has

$$\left[i\frac{\partial}{\partial t_1}+\frac{\nabla_1^2}{2m}\right]\psi(r,t)=\int dr_2\,\psi^+(r_2 t)\,v(|r_2-r_1|)\,\psi(r_2 t)\,\psi(r_1 t) \tag{1-44}$$

and from (1-7)

$$i\frac{\partial}{\partial t_1}[T\psi(1)\psi^+(1')] = i\delta(t_1-t_1')\,\delta(r_1-r_1') \pm \theta(t_1'-t_1)\,\psi^+(1')\,i\frac{\partial\psi(1)}{\partial t_1} + \theta(t_1-t_1')\,i\frac{\partial\psi(1)}{\partial t_1}\,\psi^+(1'), \tag{1-45}$$

Substituting from (1-44) into (1-45)

$$\begin{aligned}\left[i\frac{\partial}{\partial t_1}+\frac{\nabla_1^2}{2m}\right][T\psi(1)\psi(1')] &= i\,\delta(1-1') \\ &\pm\Big[\theta(t_1'-t_1)\int dr_2\,\psi^+(1')\,\psi^+(\mathbf{2}^+)\,v(|r_2-r_1|)\,\psi(2)\,\psi(1) \\ &\pm\theta(t_1-t_1')\int dr_2\,\psi^+(\mathbf{2}^+)\,v(|r_2-r_1|)\,\psi(2)\,\psi(1)\,\psi^+(1')\Big]_{t_2=t_1} \\ &= i\delta(1-1')\pm\int dr_2\,v(r_2-r_1)\,[T\psi(1)\,\psi(2)\,\psi^+(\mathbf{2}^+)\,\psi^+(1')]_{t_2=t_1}\end{aligned} \tag{1-46}$$

1 2 2⁺ 1'

$t_1' > t_2^+ > t_2 > t_1$ two transpositions $(\pm)^2$

$t_1' > t_1 > t_2^+ > t_2$ $t_1 = t_2$

$t_2^+ > t_2 > t_1' > t_1$ $t_1 = t_2$

$t_2^+ > t_2 > t_1 > t_1'$ one transposition $\pm$

$t_1 > t_2^+ > t_2 > t_1'$

$t_1 > t_2^+ > t_2 > t_1'$

$$\frac{1}{i} < T\psi(1)\psi(1') > = G\,(1, 1')$$

$$- < T\psi(1)\,\psi(2)\psi^{+}(2^{+})\,\psi^{+}(1') > = G_2\,(12;\, 1'2^{+})$$

$$\left[i\frac{\partial}{\partial t_1} + \frac{V_1^{\,2}}{2m}\right] G(1, 1') = \delta(1 - 1') \pm i \int dr_2\, v(|r_1 - r_2|)\, G_2\,(12;\, 1'2^{+})\Big|_{t_2 = t_1} \qquad (1\text{-}47)$$

Equation (1-47) determines G in terms of G_2. It is in general impossible to know G_2 exactly. We shall find G by making approximations for G_2 in the equation of motion.

However, even if G_2 were precisely known, (1-47) would not be sufficient to determine G unambiguously. This equation is first order differential equation in time, and thus a single supplementary boundary condition is required to fix the solution precisely. The necessary boundary condition is, of course, condition (1-13)

$$G(1, 1')\Big|_{t_1 = 0} = \pm e^{\beta\mu}\, G(1, 1')\Big|_{t_1 = -i\beta} \qquad (1\text{-}48)$$

A very natural way of taking this quasi-periodic boundary condition into account is to express G as a Fourier series, which we write in momentum space as

$$G(p, t - t') = \frac{1}{-i\beta} \sum_{\nu} e^{-iz_\nu (t-t')}\, G(p, z_\nu) \quad \text{for} \quad \begin{matrix} 0 \le it \le \beta \\ 0 \le it' \le \beta \end{matrix} \qquad (1\text{-}49)$$

$$e^{-iz_\nu(0-t')} = \pm e^{\beta\mu}\, e^{-iz_\nu(-i\beta - t')}$$

$$1 = \pm e^{\beta\mu - \beta z_\nu}$$

$$\beta\mu - \beta z_\nu = \pm \pi\nu i \qquad \begin{matrix} \nu\text{: even for B. E.} \\ \nu\text{: odd for F. D.} \end{matrix}$$

$$z_\nu = \mu \mp \frac{\pi\nu i}{\beta} = \mu \pm \frac{\pi\nu}{-i\beta}$$

To relate G to A we invert the Fourier series (1-49)

$$G(p, z_\nu) = \int_0^{-i\beta} dt\, e^{i[\frac{\pi\nu}{-i\beta}+\mu](t-t')} \quad G(p, t-t')$$

This integral must be independent of t' and is most simply evaluated by taking t' = 0. Then

$$G(p, t) = G^{>}(p, t) = \int_{-\infty}^{\infty} \frac{d\omega}{2\pi i} e^{-i\omega t} \frac{A(p, \omega)}{1 \mp e^{-\beta(\omega-\mu)}} \qquad \text{(c. f. 1-18)}$$

and we find

$$G(p, z_\nu) = \int_{-\infty}^{\infty} \frac{d\omega}{2\pi i} \int_0^{-i\beta} dt \left[e^{i[\frac{\pi\nu}{-i\beta}+\mu-\omega]t}\right] \frac{A(p, \omega)}{1 \mp e^{-\beta(\omega-\mu)}}$$

$$= \int_{-\infty}^{\infty} \frac{d\omega}{2\pi} \cdot \frac{A(p, \omega)}{z_\nu - \omega}, \quad z_\nu = \frac{\pi\nu}{-i\beta} + \mu \qquad (1\text{-}50)$$

Thus the Fourier coefficient is just the analytic function

$$G(p, z) = \int \frac{d\omega}{2\pi} \frac{A(p, \omega)}{z - \omega} \qquad \text{Hilbert's transform.} \qquad (1\text{-}51)$$

evaluated at $z = z_\nu = \frac{\pi\nu}{-i\beta} + \mu$.

(1-51) is the analytic continuation of the function (1-50) defined on the points $z = (\pi\nu/-i\beta) + \mu$ The $A(p, \omega)$ is given by the discontinuity of $G(p, z)$ across the real axis, i. e.,

$$A(p, \omega) - i[G(p, \omega + i\epsilon) - G(p, \omega - i\epsilon)] \quad \epsilon \to 0 \qquad (1\text{-}52)$$

since

$$\frac{1}{\omega - \omega' + i\epsilon} = P\frac{1}{\omega - \omega'} - \pi i \delta(\omega - \omega')$$

(Inverse of Hilbert transform)

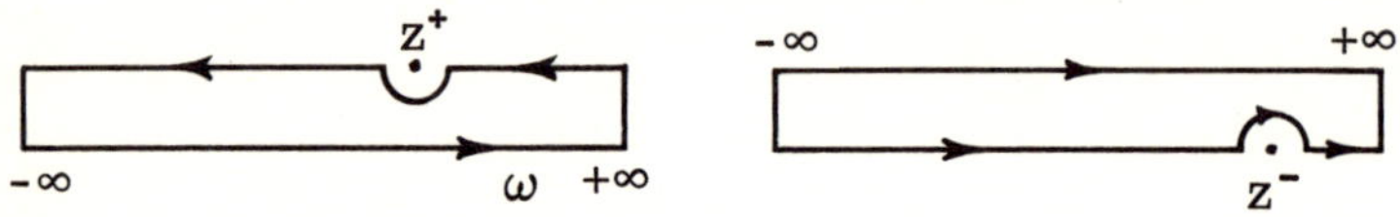

$$G(p,\ z^+) = \frac{1}{2\pi}\ P\int \frac{A(p,\ \omega)}{\omega - z^+}\,d\omega - \frac{\pi i}{2\pi}A(p,\ z^+)$$

$$G(p,\ z^-) = -\frac{1}{2\pi}\ P\int \frac{A(p,\ \omega)}{\omega - z^-}\,d\omega + \frac{\pi i}{2\pi}A(p,\ z^-)$$

$$z^+ = \omega_0 + i\epsilon$$

$$z^- = \omega_0 - i\epsilon$$

$$iA(p,\ \omega_0) = \lim_{\epsilon\to 0}\left[G(p,\ \omega_0 - i\epsilon) - G(p,\ \omega_0 + i\epsilon)\right]$$

$$A(p,\ \omega) = \lim_{\epsilon\to 0} i\left[G(p,\ \omega_0 + i\epsilon) - G(p,\ \omega_0 - i\epsilon)\right]$$

1. 5. FREE PARTICLES

$\nu = 0.$

$$\left(i\frac{\partial}{\partial t_1} + \frac{\nabla_1^2}{2m}\right) G(1,\ 1') = \delta(1 - 1') \tag{1-53}$$

Multiply this by

$$\exp[-i\,p\cdot(r_1 - r_1') + i\,z_\nu(t_1 - t_1')],\quad z_\nu = \frac{\pi\nu}{-i\beta} + \mu$$

and integrate over all r_1 and all t_1 in the interval 0 to $-i\beta$. Then (1-53) becomes after partial integrations

$$\left(z_\nu - \frac{p^2}{2m}\right) G(p,\ z_\nu) = 1$$

Therefore

$$G(p,\ z_\nu) = \frac{1}{z_\nu - (p^2/2m)} \tag{1-54}$$

The analytic continuation of this formula is

$$G(p,\ z) = \frac{1}{z-(p^2/2m)}$$

$$A(p,\ \omega) = \lim_{\epsilon\to 0} i[\,G(p,\ \omega + i\epsilon) - G(p,\ \omega - i\epsilon)\,]$$

$$= \lim_{\epsilon\to 0} i\left[\frac{1}{\omega-(p^2/2m) + i\epsilon} - \frac{1}{\omega-(p^2/2m) - i\epsilon}\right]$$

$$\boxed{P\frac{1}{\omega - (p^2/2m)} - \pi i \delta\left(\omega - \frac{p^2}{2m}\right)} \qquad \boxed{\text{same with } + \pi i \delta}$$

$$= 2\pi\ \delta\left[\omega - \frac{p^2}{2m}\right]$$

1.6. THE HARTREE APPROXIMATION

The one-particle Green's function, G(1, 1'), represents the propagation of a particle added to the medium at 1' and removed at 1. We can represent this pictorially by a line going from 1' to 1.

G(1, 1') – 1' ——>—— 1

Notice that this line represents propagation through the medium, and not free-particle propagation. Similarly,

$G_2(12{:}\ 1'2') =$

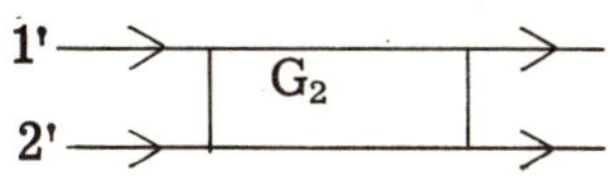

An Approximation

≅

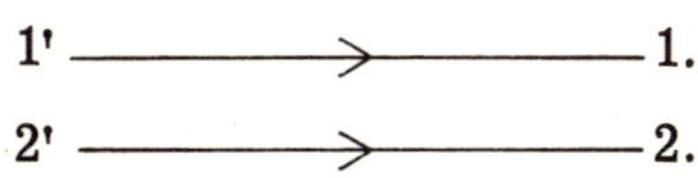

$= G(1, 1')G(2, 2')$. (1-55)

If we substitute (1-55) into the equation of motion (15-4)

$$\left[i\frac{\partial}{\partial t_1}+\frac{\nabla_1^2}{2m}\right]G(1,\ 1') = \delta(1-1') \pm i \int dr_2\, v(|r_2 - r_1|)\, G_2(1, 2; 1'2')$$

Using $G_2(12;\ 1'2') \cong G(1, 1')\, G(2, 2')$

$$\left[i\frac{\partial}{\partial t_1} + \frac{\nabla_1^2}{2m} \mp i\int dr_2\, v(r_2 - r_1)\, G(2, 2^+)\right] \ G(1, 1')$$

$$= \left[i\ \frac{\partial}{\partial t_1} + \frac{\nabla_1^2}{2m} - \int dr_2\, v(r_2 - r_1) < n(r_2) >\right] \ G(1, 1')$$

$$= \delta(1 - 1') \qquad (1\text{-}56)$$

$$G(2, 2^+) = \frac{1}{i} < T\psi(2)\, \psi^+(2^+) >$$

$$= \pm \frac{1}{i} < \psi^+(2^+)\psi(2) >$$

$$= \pm\frac{1}{i} < n(r_2) >$$

$$v(r_1) = \int dr_2\ v(r_1 - r_2)\ < n(r_2) > \qquad (1\text{-}57)$$

The self-consistent Hartree field.

For a translationally invariant system. $<n(r_2)>$ is independent of the position r_2.

Letting $v = \int dr\ v(r)$

$$v(r) = nv$$

Then, by just the same procedure as in the free-particle case, we find from (1-54) for the Fourier coefficient:

$$\left[z_\nu - \frac{p^2}{2m} - nv\right] G(p,\ z_\nu) = 1$$

The analytic continuation gives

$$G(p,\ z) = \frac{1}{z - (p^2/2m) - nv} \qquad (1\text{-}58)$$

so that in the Hartree approximation

$$A(p,\ \omega) = 2\pi\ \delta[\omega - (p^2/2m) - nv] \qquad (1\text{-}59)$$

The density of particles

$$n = \pm i\, G^{<}(rt,\ rt) = \int \frac{dp}{(2\pi)^3} \frac{d\omega}{2\pi} A(p, \omega) f(\omega) \tag{1-60}$$

which for the Hartree approximation becomes

$$n = \int \frac{dp}{(2\pi)^3} \frac{d\omega}{2\pi} \frac{1}{e^{\beta[(p^2/2m)+nv-\mu]} \mp 1} \tag{1-61}$$

The energy per unit volume using (12-4) and 20-2

$$\frac{\langle H \rangle}{\Omega} = \int \frac{dp}{(2\pi)^3} \left(\frac{p^2}{2m} + \frac{nv}{2}\right) \frac{1}{e^{\beta[(p^2/2m)+nv-\mu]} \mp 1}$$

$$= \frac{1}{2} n^2 v + \int \frac{dp}{(2\pi)^3} \frac{p^2/2m}{e^{\beta[(p^2/2m)+nv-\mu]} \mp 1} \tag{1-62}$$

The equation of state

$$dP = n\, d\mu$$

In the low density limit $\beta\mu \to -\infty$

$$n = e^{\beta(\mu - nv)} \int \frac{dp}{(2\pi)^3} e^{-\beta(p^2/2m)}$$

$$dn = \beta n\, (d\mu - v\, dn)$$

$$dP = \frac{1}{\beta} dn + v n\, dn = kT\, dn + \frac{1}{2} v\, d(n^2)$$

Since at n = 0 the pressure vanishes, we find

$$P - \frac{1}{2} n^2 v = nkT$$

1.7. THE HARTREE-FOCK APPROXIMATION

The Hartree approximation for the two-particle Green's function does not take into account the identity of the particles. Since the particles are identical, we cannot distinguish processes in which the particle added at 1' appears at 1 from processes in which it appears at 2. To include this possibility of exchange, we can write

$G_2(1\,2;\, 1'2') =$ 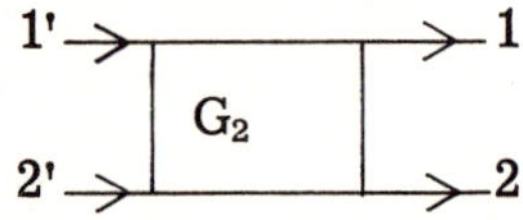

$=$ 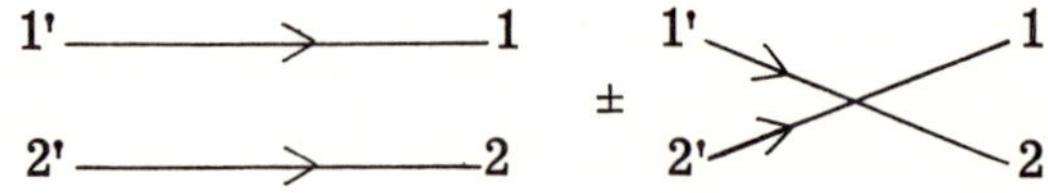

$$= G(1, 1')\, G(2, 2') \pm G(1, 2')\, G(2, 1') \qquad (21\text{-}1)$$

This approximation to G_2 leads to the Hartree-Fock approximation.

The approximate equation for G resulting from substituting (1-63) into the equation of motion takes the form

$$\left(i\frac{\partial}{\partial t_1} + \frac{\nabla_1^2}{2m}\right) G(1, 1') + \int dr_2 \langle r_1 | v | r_2 \rangle G(2,1) \Big|_{t_2 = t_1} = \delta(1 - 1') \qquad (1\text{-}63)$$

where

$$\langle r_1 | v | r_2 \rangle = \delta(r_1 - r_2) \int dr_3\, v(r_1 - r_3) \langle n(r_3) \rangle + i\, v(r_1 - r_2)\, G^{<}(1, 2) \Big|_{t_2 = t_1} \qquad (1\text{-}64)$$

Again $\langle r_1 | v | r_2 \rangle$ has the interpretation of an average, self-consistent field through which the particles move.

In the case of a translationally invariant system, we can Fourier transform (1-63) and (1-65) to obtain

$$\left[i \frac{\partial}{\partial t_1} - E(p)\right] G(p,\ t_1 - t_1') = \delta(t_1 - t_1') \qquad (1\text{-}65)$$

and

$$E(p) = \frac{p^2}{2m} + n v \pm \int \frac{dp'}{(2\pi)^3}\, v(p - p') \langle n(p') \rangle \qquad (1\text{-}66)$$

where $v(p) = \int dr e^{-ip \cdot r} v(r)$ is the Fourier transform of the potential $v(r)$. Just as before

$$A(p, \omega) = 2\pi\ \delta(\omega - E(p)) \qquad (1\text{-}68)$$

so that

$$\langle n(p) \rangle = f(E(p)) = \frac{1}{e^{\beta[E(p) - \mu]} \mp 1} \qquad (1\text{-}69)$$

The Hartree-Fock single-particle energy E(p) must then be obtained as the solution of (1-67) and (1-69).

REFERENCES

T. Matsubara, Progr. Theoret. Phys. (Kyoto), **14**, 351 (1955).

P. C. Martin and J. Schwinger, Phys. Rev , **115**, 1342 (1959).

D. N. Zubarev, Usekhi Fiz. Nauk, **71**, 71 (1960)] translation Soviet Phys. Uspekhi, **3**, 320 (1960].

Kadanoff and Baym, "Quantum Statistical Mechanics" W. A. Benjamin (1962).

Bonch-Bruevich and Tyablikov, "The Green Function Method in Statistical Mechanics", North-Holland Publishing Co. (1963).

Application of Perturbation Method

Tomoyasu Tanaka

2. 1 FORMULATION OF THE PROBLEM

Thermodynamic properties of an electron gas will be discussed by the perturbation method.

The total Hamiltonian of our system is given by

$$H = H_0 + V.$$

$$H_0 = -\sum_{j=1}^{N} \frac{\hbar^2}{2m} \nabla_j^{\,2}, \qquad V = \sum_{i>j} \frac{e^2}{|r_i - r_j|}. \tag{2-1}$$

All thermodynamic properties of the system can be derived from the partition function Z, which is defined by

$$Z = \sum_n e^{-\beta E_n} = \sum_n e^{-\beta E_n} \int \underbrace{\psi_n^*(r)\psi_n(r)\,dr}_{\text{normalized to unity}}$$

$$= \int K(r, \beta; r', \beta')\Big|_{\substack{r'=r \\ \beta'=0}} dr \tag{2-2}$$

where

$$K(r, \beta; r', \beta') = \sum_n \psi_n(r)\psi_n^*(r')e^{-E_n(\beta-\beta')} \qquad (2\text{-}3)$$

with $\beta > \beta'$.
One can immediately see that the function $K(r, \beta; r', \beta')$ is the Green's function of the Bloch equation.

$$\text{Bloch equation } \frac{\partial \rho}{\partial \beta} = -H\rho \qquad (2\text{-}4)$$

[Schrödinger equation

$$i\frac{\partial \psi}{\partial t} = H\rho, \qquad t \to -i\beta \qquad \to \text{ Bloch equation]}$$

The associated Green's function is defined as the solution of

$$\left[\frac{\partial}{\partial \beta} + H(r)\right] K(r, \beta; r', \beta') = \delta(\beta - \beta')\ \delta(r - r'). \quad (2\text{-}5)$$

$K(r, \beta; r', \beta')$ satisfies the boundary conditions

$$K = 0, \qquad r \to \infty, \qquad K = 0 \text{ for } \beta - \beta' < 0$$

$K(r, \beta; r', \beta')$ is also called the propagator.

Derivation of (2-3) from (2-5) [Morse & Feshbach, Ch. 7]

The Laplace transform of K is introduced by

$$K_L(r, E) = \int_0^\infty e^{-\beta E} K(r, \beta)\, d\beta$$

Then

$$E K_L(r, E) = -\int_0^\infty K(r, \beta) \frac{\partial e^{-\beta E}}{\partial \beta} d\beta$$

$$= -K(r, \beta) e^{-\beta E}\Big|_0^\infty + \int_0^\infty e^{-\beta E} \frac{\partial K}{\partial \beta}\, d\beta$$

$$= -H(r)\ K_L(r, E) + e^{-\beta' E}\ \delta(r_2 - r_1)$$

or

$$[H(r) + E]\, K_L(r, E) = e^{-\beta' E}\, \delta(r_2 - r_1) \tag{2-6}$$

Now $K_L(r, E)$ is expanded into series in terms of an orthonormal set $\{\psi_n\}$

$$K_L(r, E) = \sum_n a_n \psi_n(r), \tag{2-7}$$

where

$$H \psi_n(r) = E_n \psi_n(r) \quad [\text{Schrödinger equation}] \tag{2-8}$$

Introducing (2-7) into (2-6)

$$\sum_n (E + E_n)\, a_n \psi_n(r) = e^{-\beta' E}\, \delta(r - r')$$

By the orthogonality condition

$$a_n = \frac{e^{-\beta' E}}{E + E_n}\, \psi_n^*(r')$$

$$\therefore K_L(r, E) = \sum_n \frac{e^{-\beta' E}}{E + E_n}\, \psi_n(r)\, \psi_n^*(r') \tag{2-9}$$

By the inverse Laplace transform

$$K(r, \beta; r', \beta') = \frac{1}{2\pi i} \sum_n \int_{c-i\infty}^{c+i\infty} \frac{e^{(\beta-\beta')E}}{E + E_n}\, dE.\psi_n(r)\, \psi_n^*(r')$$

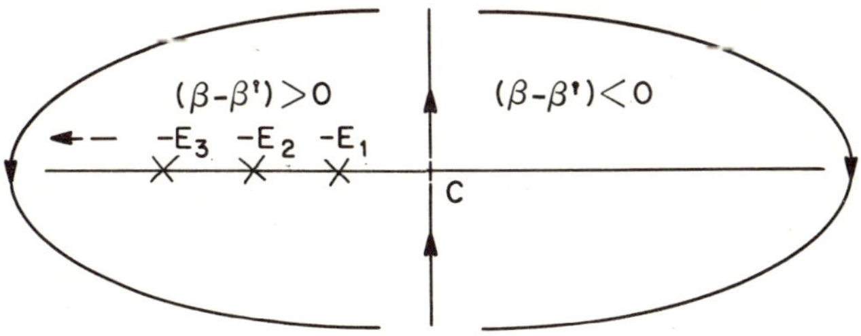

The path of integration should be chosen so that the boundary conditions are satisfied. This is done by choosing c to the right of all the poles

$$-E_1, \ -E_2, \ -E_3, \ \ldots$$

When $\beta - \beta' > 0$, $\quad e^{(\beta-\beta')E} \to 0$, Re $E \to -\infty$
The contour can be closed in the left half plane contributing

$$K(r, \beta; r', \beta') = \sum_n \psi_n(r)\psi_n^*(r')e^{-(\beta-\beta')E_n} \tag{2-10}$$

When $\beta - \beta' < 0$, $e^{(\beta-\beta')E} \to 0$, $\quad$ Re $E \to +\infty$
The contour is closed in the right half plane and

$$K(r, \beta; r', \beta') = 0, \quad (\beta - \beta') < 0. \tag{2-11}$$

(2-2), (2-10) are the basic equations of our problem.

It is, in general, impossible to solve (2-5) exactly, and the method of perturbation is introduced.

(2-5) can be written as

$$\left[\frac{\partial}{\partial\beta_2} + H_0(r_2) + V(r_2)\right] K(r_2\ \beta_2; r_1\ \beta_1) = \delta(\beta_2 - \beta_1)\,\delta(r_2 - r_1)$$

or

$$\left[\frac{\partial}{\partial\beta_2} + H_0(r_2)\right] K(r_2\beta_2; r_1\beta_1) = \delta(\beta_2 - \beta_1)\,\delta(r_2 - r_1) - \\ -V(r_2)K(r_2\beta_2; r_1\beta_1)$$

This equation, regarded as an inhomogeneous equation is solved formally

$$K(r_2\ \beta_2; r_1\ \beta_1 = \int dr_3 \int_0^{\beta_2} d\beta_3 K_0(r_2\beta_2; r_3\beta_3) \\ \times \left[\delta(\beta_3 - \beta_1)\,\delta(r_3 - r_1) - V(r_3)K(r_3\ \beta_3; r_1\beta_1)\right] \tag{2-12}$$

Where K_0 is the Green's function associated with the homogeneous part, i. e.

$$\left[\frac{\partial}{\partial\beta_2} + H_0(2)\right] K_0(r_2\beta_2; r_1\beta_1) = \delta(\beta_2 - \beta_1)\delta(r_2 - r_1). \tag{2-13}$$

(2-12) becomes

$$K(r_2\,\beta_2\,;r_1\,\beta_1) = K_0(r_2\,\beta_2\,;r_1\,\beta_1) - \int dr_3 \int_0^{\beta_2} K_0(r_2\,\beta_2\,;r_3\,\beta_3)$$

$$V(r_3)K(r_3\beta_3\,;r_1\beta_1) \quad (2\text{-}14)$$

This is an integral equation for K and the solution is found by the method of successive substitution.

$$K(2,\ 1) = K_0(2,1) - \int d(3) \int_{\beta_1}^{\beta_2} d\beta_3 K_0(2,\ 3)\,V(3)\,K_0(3,\ 1)$$
$$+ \int d(3) \int_{\beta_1}^{\beta_2} d\beta_3 K_0(2,3)\,V(3) \int d(4)$$
$$\int_{\beta_1}^{\beta_3} d\beta_4 K_0(3,\ 4)\,V(4)\,K_0(4,\ 1) - + \ldots \quad (2\text{-}15)$$

2.2 THE GREEN's FUNCTION FOR THE FREE PARTICLE HAMILTONIAN

If we choose H_0 to be the free particle Hamiltonian, it is easy to prove that

$$K_0(r,\beta;r'\beta') = K_0^{(N)}(r_1, r_2, \ldots, r_N, \beta; r_1', r_2', \ldots, r_N', \beta') \quad (2\text{-}16)$$
$$= \prod_{j=1}^{N} K_0(r_j,\beta;\, r_j',\, \beta',) \quad \text{[Max Boltz]}$$

where

$$K_0(r\,\beta;\,r'\,\beta') = \frac{\exp[-\frac{1}{2}\,m(r-r')^2/\hbar^2(\beta-\beta')]}{[(2\hbar^2\,\pi/m)\,(\beta-\beta')]^{3/2}} \quad (2\text{-}17)$$

This is obtained from the free particle wave function

$\frac{1}{\sqrt{V}}\,e^{ik\cdot r}$ and the energy eigenvalue: $\frac{\hbar^2 k^2}{2m}$

We have

$$K_0(r,\,\beta\,;r'\,\beta') = \sum_k \frac{1}{V}\,e^{ik\cdot(r-r')}\; e^{-\frac{\hbar^2 k^2}{2m}(\beta-\beta')} \quad (2\text{-}18)$$

$k = \frac{2\pi n}{V^{\frac{1}{3}}}$, The number of states between $k \sim k + dk$

(where $n = n_x,\ n_y,\ n_z$ each 0, ±1, ±2, ...)

Replacing the summation by the $\frac{V}{(2\pi)^3}\,dk$ integration

$$K_0(r\,\beta;\,r'\,\beta') = \frac{1}{(2\pi)^3}\int e^{-\frac{\hbar^2 k^2}{2m}(\beta-\beta') + ik\cdot(r-r')}\,dk \tag{2-19}$$

$$= \frac{\exp[-\frac{1}{2}m(r-r')^2/\hbar^2(\beta-\beta')]}{[(2\hbar^2\pi/m)(\beta-\beta')]^{3/2}}$$

$$\int e^{-ak^2}\,dk = \left(\frac{\pi}{a}\right)^{3/2} \qquad \int e^{-a(k^2-2ik\cdot y)}\,dk = \left(\frac{\pi}{a}\right)^{3/2} e^{-ay^2}$$

$$k \rightarrow k - iy, \quad k^2 \rightarrow k^2 - 2iky - y^2$$

$$a = \frac{\hbar^2}{2m}(\beta-\beta'); \quad 2ay = r - r'; \; ay^2 = \frac{(r\,r')^2}{4a}$$

Another example for which one can calculate the Green's function in closed form is the harmonic oscillator. The answer is identical with (7- 13) in the first set of lectures, replacing β by $\beta - \beta'$. See also Elliot W. Montroll: La théorie des gaz neutres et ionisé (sec. 1. 2b).

Fermi-Dirac case

The antisymmetric normalized eigenfunction is

$$\Psi_N(r) = \frac{1}{\sqrt{N!}},\ \frac{1}{\sqrt{V^N}} \begin{vmatrix} e^{ik_1\cdot r_1} & e^{ik_1\cdot r_2} & \dots & e^{ik_1\cdot r_N} \\ e^{ik_2\cdot r_1} & e^{ik_2\cdot r_2} & \dots & e^{ik_2\cdot r_N} \\ \cdot & & & \\ \cdot & & & \\ e^{ik_N\cdot r_1} & e^{ik_N\cdot r_2} & \dots & e^{ik_N\cdot r_N} \end{vmatrix} \begin{matrix} k_1 \\ k_2 \\ \\ \\ k_N \end{matrix}$$

$$\begin{matrix} r_1 & r_2 & & r_N \end{matrix} \tag{2-20}$$

$$= \frac{1}{\sqrt{N!}}\frac{1}{\sqrt{V^N}} \sum_P (-1)^p\, P \exp i[k_1\cdot r_1 + k_2\cdot r_2 + \dots + k_N\cdot r_N] \tag{2-21}$$

$$K_0(r, \beta; r', \beta') = \sum_{\{k\}} \Psi_{N(k)}(r)\Psi^*_{N(k)}(r')\, e^{-E_k(\beta-\beta')}$$

$$E_k = \sum_{j=1}^{N} \frac{\hbar^2 k_j^2}{2m}$$

$$= \frac{1}{N!}\frac{1}{V^N}\sum_{\{k\}}\sum_{P}\sum_{Q'}(-1)^{p+q} PQ' \prod_{j=1}^{N} \exp\left[ik_j\cdot(r_j - r_j') - \frac{\hbar^2 k_j^2}{2m}(\beta - \beta')\right]$$

$$= \left(\frac{1}{N!}\right)^2 \sum_{P}\sum_{Q}(-1)^{p+q} PQ' \prod_{j=1}^{N} K_0(r_j, \beta; r_j'\, \beta') \qquad (2\text{-}22)$$

The multiple summation over $\{k\}$ was for one chosen order of k' s, if we drop this restriction a factor $1/N!$ is necessary.

$$Q' = Q_1' P' \qquad \sum_{Q} Q' = \sum_{Q_1} Q_1' P'$$

$$q = q_1 + p$$

$$K_0(r, \beta; r'\, \beta) = \left(\frac{1}{N!}\right)^2 \sum_{P}\sum_{Q_1'}(-1)^{2p+q_1} PQ_1' P' \prod_{j=1}^{N} K_0(r_j\, \beta; r_j'\, \beta')$$

$$= \frac{1}{N!}\sum_{Q_1'}(-1)^{q_1} Q_1' \prod_{j=1}^{N} K_0(r_j\, \beta; r_j'\, \beta')$$

$$= \frac{1}{N!}\begin{vmatrix} K_0(r_1\, \beta; r_1'\, \beta') & K_0(r_1\beta; r_2'\, \beta') & \dots & K_0(r_1\, \beta; r_N'\, \beta') \\ K_0(r_2\, \beta; r_1'\, \beta') & K_0(r_2\beta; r_2'\, \beta') & \dots & K_0(r_2\, \beta; r_N'\, \beta') \\ \\ \\ K_0(r_N\beta; r_1'\, \beta') & K_0(r_N\beta; r_2'\, \beta') & \dots & K_0(r_N\beta; r_N'\, \beta') \end{vmatrix} \qquad (2\text{-}23)$$

$$= \frac{1}{N!}\, \det{}_N K_0(r, \beta; r'\, \beta')$$

The grand partition function Z_G is obtained from the propagator $K_0(r, \beta; r', \beta')$ first by setting $r' = r$, $\beta' = 0$, then integrating over the coordinates and finally summing over all numbers of particles.

$$Z_G = \sum_{N=0}^{\infty} z^N \int_V \cdots \int \det{}_N K_0(r, \beta; r, 0)\, dr_1 dr_2 \ldots dr_N \tag{2-24}$$

$$= \sum_{N=0}^{\infty} z^N \sum_{\{r_t\}} \prod_t \frac{(-1)^{(t+1) r_t} (A_t/t)^{r_t}}{r_t\,!} \tag{2-25}$$

$$\left\{ \sum_{t=1}^{N} t\, r_t = N \right\}$$

where

$$A_t = \int_V \cdots \int K_0(r_1 \beta; r_t, 0) K_0(r_t \beta; r_{t-1} 0) \cdots K_0(r_3 \beta; r_2 0) .$$
$$K_0(r_2 \beta; r_1 0)\, dr_1 \cdots dr_t \tag{2-26}$$

(2-25) follows from (2-24) by using the properties of a determinant.
From (2-23)

$$K_0(r \beta; r' \beta') = \frac{1}{N!} \sum_Q (-1)^q Q \prod_{j=1}^{N} K_0(r_j \beta; r_j' \beta')$$

Q is a permutation operator which operates on the N indices of r_j''s (r_1', r_2', r_3', ..., r_N') and the summation is taken over N! permutations.

Since any permutation of N indices can be represented as a product of independent cyclic permutations the structure of integrals A_t is clear.

EXAMPLE:

$$Q = \begin{pmatrix} 2 & 3 & 1 & 5 & 4 \\ 1 & 2 & 3 & 4 & 5 \end{pmatrix}$$

$$= (1\ 2\ 3)\ (4\ 5)$$

(4 5): cycle of 2 indices
(1 2 3): cycle of 3 indices

$$K_0(r_1\,\beta; r_2'\,0)\ K_0(r_2\,\beta; r_3'\,0)\,K_0(r_3\,\beta; r_1'\,0) \vdots\ K_0(r_4\,\beta; r_5'\,0)$$
$$K_0(r_5\,\beta; r_4'\,0)$$

A cycle of t indices is obtained as a result of (t + 1) transpositions of 2 indices:

$$(1\,2\,3) \to \begin{matrix} 2\,3\,1 \\ \uparrow 2\,1\,3 \\ \uparrow 1\,2\,3 \end{matrix} \to (13)\,(12),\ (-1)^q \to \prod_{t=1}^{N} (-1)^{(t+1)\,r_t}$$

Number of ways of obtaining r_t cycles of t indices (t = 1, 2, . . ., N) is

$$\prod_{t=1}^{N} \frac{N!}{(r_t!)\,(t!)^{r_t}}\,[\,(t-1)\,!\,]^{r_t} \qquad (2\text{-}27)$$

Now our sum of products of A_t's is exactly

$$Z_G = \exp \sum_{t=1}^{\infty} (-1)^{t+1}\ z^t\ \left(\frac{A_t}{t}\right) \qquad (2\text{-}28)$$

so that

$$\log Z_G = \sum_{t=1}^{\infty} (-1)^{t+1}\ z^t\ (A_t/t) \qquad (2\text{-}29)$$

The integrals A_t's can be interpreted in the following way:- Because of Pauli's exclusion principle there exists an effective interaction (which may be called the statistical interaction) between particles, and this gives rise to various cluster integrals. But in an ideal system those clusters are rings in configuration space. The cyclic or ring nature of the cluster integrals has been noted by Kahn and Uhlenbeck (1939). Montroll and Ward considered these clusters in both reciprocal temperature and configuration spaces and called them the torons.

$$\int K_0(r_1\beta; r_t\,0)\,K(r_t\,\beta; r_{t-1},\,0) \ldots K_0(r_3\beta; r_2\,0)$$
$$K_0(r_2\beta; r_1\,0)\,dr_1 \cdots dr_t$$

a ring cluster

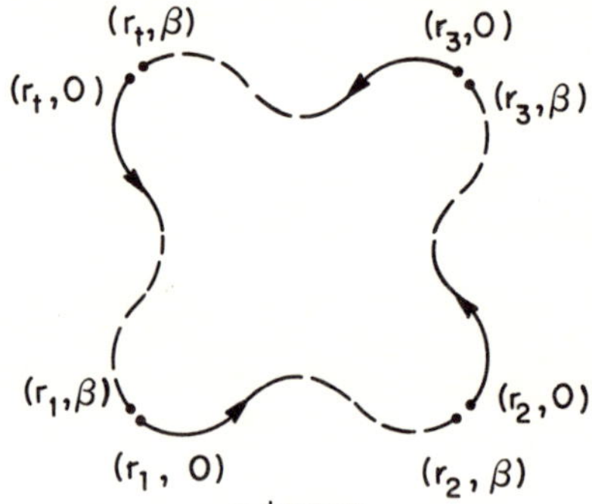

a toron

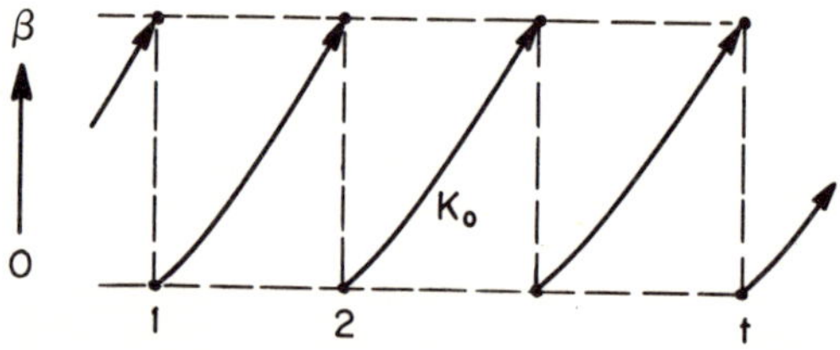

The explicit calculation of the A's is done by employing the Fourier transforms, [(2-18) or (2-19)], of the free particle propagators.

$$A_t(\beta) = \int_V \cdots \int d\mathbf{r}_1 \cdots d\mathbf{r}_t \prod_{j=1}^{t} K_0(\mathbf{r}_{j+1}, \beta; \mathbf{r}_j, 0),$$

$$\mathbf{r}_{t+1} = \mathbf{r}_1 .$$

$$= \sum_{\mathbf{p}_1} \cdots \sum_{\mathbf{p}_t} \frac{1}{V^t} \int \prod_{j=1}^{t} d\mathbf{r}_j \exp\Big[i \sum_{j=1}^{t} \frac{\mathbf{p}_j}{\hbar} \cdot (\mathbf{r}_{j+1} - \mathbf{r}_j)$$

$$- \sum_{j=1}^{t} \frac{p_j^2}{2m} \beta \Big], \quad \mathbf{p}_j = \hbar \mathbf{k}_j$$

$$\sum_{j=1}^{t} \mathbf{p}_j \cdot (\mathbf{r}_{j+1} - \mathbf{r}_j) = \mathbf{p}_1 \cdot (\mathbf{r}_2 - \mathbf{r}_1) + \mathbf{p}_2 \cdot (\mathbf{r}_3 - \mathbf{r}_1) + \ldots + \mathbf{p}_t \cdot (\mathbf{r}_1 - \mathbf{r}_t)$$

$$= \mathbf{r}_1 \cdot (\mathbf{p}_t - \mathbf{p}_1) + \mathbf{r}_2 \cdot (\mathbf{p}_1 - \mathbf{p}_2) + \ldots + \mathbf{r}_t \cdot (\mathbf{p}_{t-1} - \mathbf{p}_t)$$

$$\frac{1}{V} \int e^{i\mathbf{r}_1 \cdot (\mathbf{p}_t - \mathbf{p}_1)} \, d\mathbf{r}_1 = \delta(\mathbf{p}_{t_1}, \mathbf{p}_1); \; \frac{1}{V} \int e^{i\mathbf{r}_2 \cdot (\mathbf{p}_1 - \mathbf{p}_2)} \, d\mathbf{r}_2 = \delta(\mathbf{p}_1, \mathbf{p}_2);$$

etc.

$$\sum_{p_1} \cdots \sum_{p_t} = \sum_p$$

$$A_t(\beta) = \sum_p \exp\left(-\frac{tp^2}{2m}\beta\right)$$

$$= \frac{V}{(2\pi)^3} \int \frac{dp}{\hbar^3} \exp\left(-\frac{t\beta}{2m} p^2\right)$$

$$= \frac{V}{(2\pi\hbar)^3} \int_0^\infty 4\pi p^2 \exp(-t\beta p^2/2m) \qquad (2\text{-}30)$$

Equation (2-29) then yields the well-known Fermion grand partition function

$$\log Z = \frac{V}{(2\pi\hbar)^3} \int_0^\infty 4\pi p^2 \sum_{t=1}^\infty (-1)^{t+1} z^t \frac{e^{-t\beta p^2/2m}}{t}$$

$$= \frac{4\pi V}{(2\pi\hbar)^3} \int_0^\infty p^2 \log\left\{1 + z\, e^{-p^2/2mkT}\right\} dp \qquad (2\text{-}31)$$

Note of the editor:

The calculation of the grand partition function for an ideal Fermion gas can be performed in a shorter way (compare Montroll in Lectures in Theoretical Physics III, pages 273-274) by recognizing the fact the quantum mechanical Feynman graphs differ from the classical only by the occurrence of "crossovers." In the Toron language classical particles move only once around the torus, quantum mechanical particles once, twice, etc. If we have a line that moves l times around the torus we realize that it has to be "propagated" over a distance of $l\beta$. Hence, the classical exponential $\exp(-p^2\beta/2m)$ is modified into:

$$e^{-p^2 l\beta/2m}$$

One has to realize that a line turning around l times, in the Torus language, means in the Feynman graph language l lines and hence l particles. The total number of particles is

$$N = \sum_l l_l S_l \qquad (1)$$

where S_l is the number of lines that turn l-times around the torus.

The net result is

$$\prod_{l=1}^{N} \left\{ (-1)^{l+1} \frac{V}{(2\pi\hbar)^3} \int e^{-p^2 l\beta/2m} d^3p \right\}^{S_l}$$

where the factor -1 came in because we are dealing with fermions. The integral is the result of an unlabelled graph. This graph can be labelled in l different ways (the number of cyclic permutations of l numbers). Hence the contribution from one integral is

$$\frac{(-1)^{l+1}}{l} \frac{V}{(2\pi\hbar)^3} \int e^{-p^2 l\beta/2m} d^3p = b_l$$

The total contribution is

$$Z_N = \sum_{S_l} \prod_l \frac{1}{N!} \frac{N!}{S_1!\, S_2! \ldots} b_l^{S_l}$$

The first combinatorial factor stems from the fact that all the N p_x, p_y, p_z - integrations include the permutation of particles and we want to include only one configuration. The second factor is the number of ways we can divide N in the numbers S_l such that (1) is fulfilled. It is the generalization of the binomial coefficient as used in the elementary Maxwell-Boltzman derivation.

In the summation over S_l, which is a multiple summation over S_1, S_2, S_3, etc., one is "hindered" by the restriction (1) and in order to avoid this the grand partition function is introduced.

$$Z_G = \sum_N z^N Z_N = \exp \sum_{l=1}^{\infty} b_l z^l$$

After substituting the integral and interchanging integral and summation, one can perform the summation. One finds again (2-31).

2.3 GRAND PARTITION FUNCTION OF INTERACTING PARTICLES.

An important property of the free propagators. From (2-18)

$$K_0(r_1\beta_1;\ r_2\beta_2) = \frac{1}{V}\sum_p e^{i\frac{p}{\hbar}\cdot(r_1-r_2)}\, e^{-\frac{p^2}{2m}(\beta_1-\beta_2)},\quad \beta_1 > \beta_2 > 0$$

$$\int K_0(r_1\beta_1;\ r_3\beta_3)\, K_0(r_3\beta_3;\ r_2\beta_2)\, dr_3$$

$$= \frac{1}{V^2}\sum_{p_1}\sum_{p_2} e^{-\frac{p_1^2}{2m}(\beta_1-\beta_3)-\frac{p_2^2}{2m}(\beta_3-\beta_2)} \int dr_3\, e^{i\frac{p_1}{\hbar}(r_1-r_3)+i\frac{p_2}{\hbar}(r_3-r_2)}$$

$$\frac{1}{V}\int dr_3\ {}^{ir_3(p_2-p_2)/\hbar} = \delta(p_1, p_2)$$

$$= \frac{1}{V}\sum_p e^{-\frac{p^2}{2m}(\beta_1-\beta_2)}\, e^{ip\cdot(r_1-r_2)/\hbar} \tag{2-32}$$

$$= K_0(r_1\beta_1;\ r_2\beta_2),\qquad (\beta_1-\beta_3)+(\beta_3-\beta_2) = (\beta_1-\beta_2)$$

Now, from (2-15)

$$K(r',\ \beta;\ r,\ 0) = K_0(r',\ \beta;\ r,\ 0)$$

$$- \int_0^\beta\!\int K_0(r',\ \beta;\ r'',\ \beta'')\ V(r'')\, K_0(r''\ \beta'';\ r,\ 0)\ d\beta''\ dr''$$

$$+ \int_0^\beta\!\!\int K_0(r'\beta;\ r''\ \beta'')\, V(r'') \int_0^{\beta''}\!\!\int K_0(r'',\ \beta'';\ r'''\ \beta''')\, V(r''')$$

$$K_0(r'''\beta''';\ r,\ 0)\ d\beta'''\ d\beta'''\ dr'''\ dr''$$

$$- + \ldots . \tag{2-33}$$

In order to calculate the partition function, we set r' = r and integrate over r.

$$K_0(r, \beta; r, 0) \xrightarrow{\text{contribute}} \text{sum of products of cycles of propagators (free torons)}$$

$$K_0 V K_0 \rightarrow V(1;3)$$

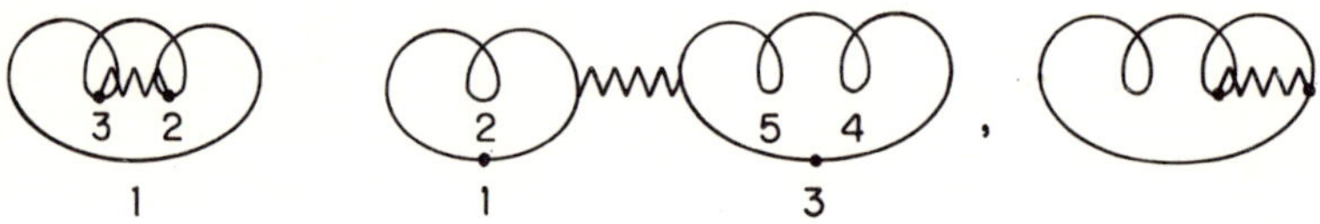

$$Z_N = N! \sum_{\{r_l\}} \prod_{l=1}^{N} \frac{b_l^{r_l}}{r_l!} \tag{2-34}$$

$$\sum_{l=1}^{N} l\, r_l = N$$

where b_l is a weighted sum of all irreducible integrals which can be constructed with l particles and all possible interactions between them.

The number of ways of constructing r_l groups (irreducible clusters) of l particles is given by

$$\frac{N!}{r_1!\, r_2! \cdots r_N! (1!)^{r_1} (2!)^{r_2} \cdots (l!)^{r_l}}$$

$$Z_G = \sum_{N=0}^{\infty} \frac{1}{N!} z^N Z_N$$

$$\log Z_G = \sum z^l b_l$$

2.4 CONTRIBUTION OF RING INTEGRALS TO GRAND PARTITION FUNCTION.

In view of the difficulties encountered in the computation of classical cluster integrals, we cannot expect to breeze through the explicit calculation of their quantum-mechanical counterparts. However, one class of cluster integrals which can be computed in detail is the ring integral in which each particle interacts with two others.

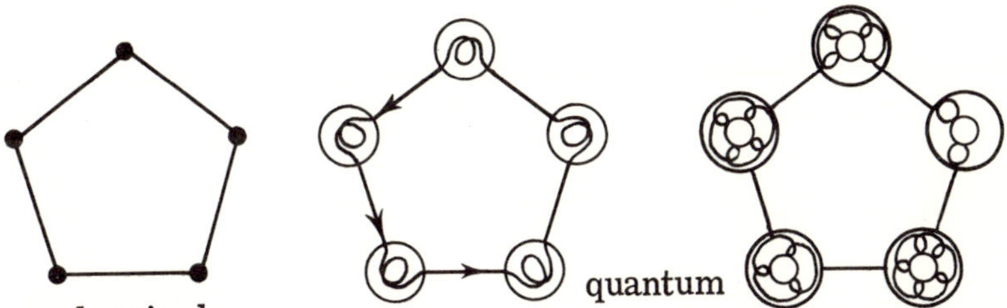

classical ring integral — Boltzmann ring* — quantum Fermion ring

We introduce the theory of ring integrals by discussing a Boltzmann gas, which is a system of interacting torons of the first order. We then generalize to a system of fermions by considering torons of all order.

As an example we consider a toron of three particles. This appears in the fourth term of perturbation expansion.

$$\int K_0(r_1^*,\ \beta;\ r_1'\ \beta_1)\ K_0(r_2^*,\ \beta;\ r_2'\ \beta_1)\ K_0(r_3^*,\ \beta;\ r_3',\ \beta_1)$$

$$\cdots\ [V(r_1' - r_2') + V(r_2' - r_3') + \underline{V(r_3' - r_1')} + \ldots]$$

$$\times K_0(r_1',\ \beta_1;\ r_1''\ \beta_2)\ K_0(r_2'\ \beta_1;\ r_2''\ \beta_2)\ K_0(r_3'\ \beta_1;\ r_3''\ \beta_2)$$

$$\cdots [V(r_1'' - r_2'') + \underline{V(r_2'' - r_3'')} + V(r_3'' - r_1'') + \cdots]$$

$$\times K_0(r_1''\ \beta_2;\ r_1'''\ \beta_3)\ K_0(r_2''\ \beta_2;\ r_2'''\ \beta_3)\ K_0(r_3''\ \beta_2;\ r_3'''\ \beta_3)$$

$$\cdots [\underline{V(r_1''' - r_2''')} + V(r_2''' - r_3''') + V(r_3''' - r_1''') + \cdots]$$

$$\times K_0(r_1'''\ \beta_3;\ r_1 0)\ K_0(r_2'''\ \beta_3;\ r_2 0)\ K_0(r_3'''\ \beta_3;\ r_3 0)$$

$$\cdots d\beta_1\, d\beta_2\, d\beta_3\, dr'\, dr''\, dr'''\ \ \delta(r^* - r)\, dr\, dr^* \tag{2-35}$$

*Diffusion effect included, statistical effect not included.

We shall concentrate our attention only to the toron formed by 1, 2, 3 particles. (Out of the 27 terms we take the underlined term as an example.)

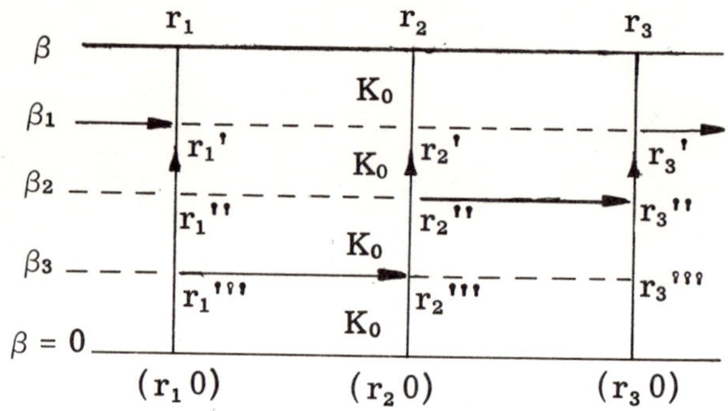

Integration over r_2, r_2'', r_2'.

$$K_0(r_2, \beta; r_2', \beta_1)\, K_0(r_2', \beta_1; r_2'' \beta_2)\, V(r_2'' - r_3'')$$

dr_2' $\downarrow$

$$K_0(r_2'' \beta_2; r_2''' \beta_3)\, K_0(r_2''' \beta_3; r_2 0)$$

$$K_0(r_2, \beta; r_2'' \beta_2) \longleftarrow dr_2$$

$$K_0(r_2''' - r_2''; \beta - \beta_2 + \beta_3)\, V(r_2'' - r_3'')\, K_0(r_2'' - r_2'''; \beta_2 - \beta_3)$$

From (2-18)

$$\frac{1}{V^2} \sum_p \sum_q \int e^{-\frac{p^2}{2m}[\beta-(\beta_2-\beta_3)] - \frac{q^2}{2m}(\beta_2-\beta_3)}\, e^{i\frac{p}{\hbar}\cdot(r_2''' - r_2'') + i\frac{q}{\hbar}\cdot(r_2'' - r_2''')}$$

$$\times V(r_2'' - r_3'')\, dr''$$

$$= \frac{1}{V^2} \sum_p \sum_q \int e^{-\frac{p^2}{2m}[\beta-(\beta_2-\beta_3)] - \frac{q^2}{2m}(\beta_2-\beta_3)}\, e^{i(p-q)\cdot(r_2''' - r_2'')/\hbar}$$

$$\times V(r_2'' - r_3'')\, dr_2''$$

Introduce: $e^{i(p-q)\cdot(r_2''' - r_3'')/\hbar} \int e^{-i(p-q)\cdot(r_2'' - r_3'')}\, V(|r_2'' - r_3''|)\, dr_2''$

and call the integral $u(p - q)$.

$$= \frac{1}{V^2} \sum_p \sum_q e^{-\frac{p^2}{2m}[\beta-(\beta_2-\beta_3)]} \times$$

$$\times\, e^{-\frac{q^2}{2m}(\beta_2-\beta_3)} \quad e^{i(p-q)\cdot(r_2'''-r_3'')/\hbar}\, u(|p-q|)$$

$$p - q \quad \rightarrow \quad -q'$$

$$q \quad \rightarrow \quad p + q'$$

$$= \frac{1}{V^2} \sum_p \sum_q e^{-\frac{p^2}{2m}[\beta-(\beta_2-\beta_3)]} e^{-\frac{(p+q)^2}{2m}(\beta_2-\beta_3)} e^{-iq\cdot(r_2'''-r_3'')/\hbar} \quad u(q) \tag{2-36}$$

The last expression is a function of relative distance between r_2''' and r_3'', and is expressed by the following diagrams:

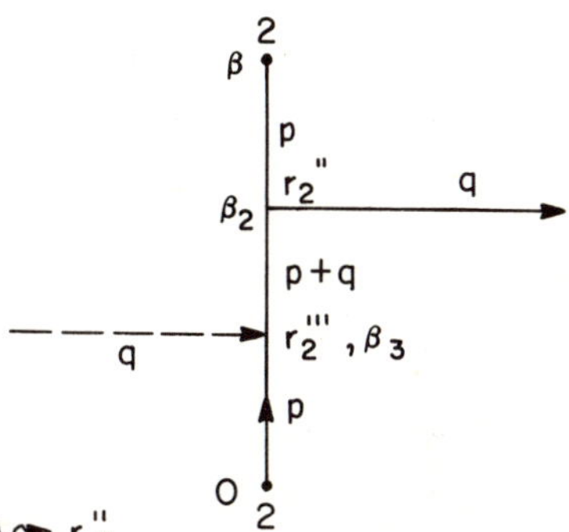

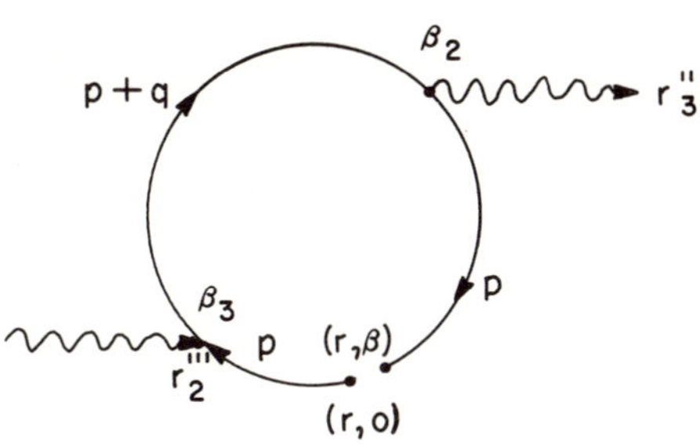

The toron is created at temperature 0 with momentum p, absorbes a quantum q which is emitted by toron # 1, at temperature β_3, emitts a quantum q at temperature β_2 and then is annihilated at temperature β. Of its entire "life"

β the toron has a momentum $(p + q)$ for the interval $\beta_2 - \beta_3$, and p for an interval $\beta - (\beta_2 - \beta_3)$.
If the sum over p is replaced by integral,

$$\sum_p \sim \frac{V}{(2\pi\hbar)^3}\, dp,$$

we obtain

$$\frac{1}{(2\pi\hbar)^{3/2}} \cdot \frac{1}{(2\pi m/\beta)^{3/2}} \frac{1}{V} \sum_q e^{-\frac{q^2\alpha}{2m\beta}(\beta-\alpha) - iq\cdot(r_2''' - r_3'')/\hbar}\, u(q). \tag{2-37}$$

where $\alpha = \beta_2 - \beta_3$

$$e^{-\frac{p^2}{2m}[\beta-(\beta_2-\beta_3)] - \frac{p^2}{2m}(\beta_2-\beta_3) - \frac{2p\cdot q}{2m}(\beta_2-\beta_3) - \frac{q^2}{2m}(\beta_2-\beta_3)}$$

$$= e^{-\frac{p^2}{2m}\beta - \frac{2p\cdot q}{2m}\alpha - \frac{q^2}{2m}\alpha}$$

$$= e^{-\frac{\beta}{2m}(p^2 + 2p\cdot q\frac{\alpha}{\beta} + \frac{\alpha^2}{\beta^2}q^2) + \frac{\alpha^2 q^2}{2m\beta} - \frac{q^2}{2m}\alpha}$$

$$(2\pi m/\beta)^{3/2} e^{-\frac{q^2}{2m\beta}\alpha(\beta-\alpha)} e^{-iq\cdot(r_2''' - r_3'')/\hbar}\, u(q). \tag{2-38}$$

The contribution of the toron No. 3 is now written down:

$$\frac{1}{(2\pi\hbar)^3} (2\pi m/\beta)^{3/2} \frac{1}{V} \sum_q e^{-\frac{q^2\alpha}{2m\beta}(\beta-\alpha) - iq\cdot(r_3'' - r_1')/\hbar} u(q). \tag{2-39}$$

$$\alpha = \beta_1 - \beta_2$$

The contribution of the toron No. 1 is, however, slightly different from the others.

$$K_0(r_1 - r_1'; \beta - \beta_1)\, K_0(r_1' - r_1''', \beta_1 - \beta_3)\, V(r_1''' - r_2''')\, K_0(r_1''' - r_1; \beta_3)$$

Integrating over r_1:

$$K_0(r_1''' - r_1'; \beta - \beta_1 + \beta_3)\, K_0(r_1' - r_1''', \beta_1 - \beta_3)\, V(r_1''' - r_2''')$$

Integrating over r_1''', we obtain

$$\frac{1}{V^2}\sum_p\sum_q e^{-\frac{p^2}{2m}[\beta-(\beta_1-\beta_3)]-\frac{q^2}{2m}(\beta_1-\beta_3)}\int e^{ip\cdot(r_1'''-r_1')/\hbar+iq\cdot(r_1'-r_1''')/\hbar}$$

$$V(r_1'''-r_2''')\,dr_1'''$$

$$=\frac{1}{V^2}\sum_p\sum_q e^{-\frac{p^2}{2m}[\beta-(\beta_1-\beta_3)]-\frac{q^2}{2m}(\beta_1-\beta_3)}e^{i(p-q)\cdot(r_2'''-r_1')/\hbar}\quad u(|p-q|)$$

Either we introduce

$$p-q=q_1,\qquad q=p-q_1$$

$$\frac{1}{V^2}\sum_p\sum_q e^{-\frac{p^2}{2m}[\beta-(\beta_1-\beta_3)]-\frac{(p-q)^2}{2m}(\beta_1-\beta_3)}e^{-iq(r_1'-r_2''')/\hbar}\quad u(q) \tag{2-40}$$

Or by introducing $p_1 = p - q$, we obtain

$$\frac{1}{V^2}\sum_p\sum_q e^{-\frac{(p+q)^2}{2m}[\beta-(\beta_1-\beta_3)]-\frac{p^2}{2m}(\beta_1-\beta_3)}e^{-iq\cdot(r_1'-r_2''')/\hbar}\quad u(q)\tag{2-41}$$

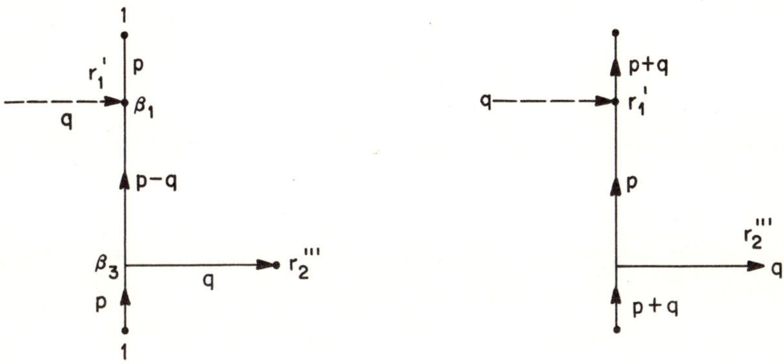

(2-40) (2-41)

However if we integrate over p, we obtain the same answer

$$\frac{1}{(2\pi\hbar)^3V}\;\frac{2\pi m^{3/2}}{\beta}\sum_q e^{-\frac{q^2\alpha'}{2m\beta}(\beta-\alpha')-iq\cdot(r_1'-r_2''')/\hbar}\quad u(q)\tag{2-42}$$

where $\alpha' = \beta_1 - \beta_3 > 0$

In this case it is more convenient to define α to be

$$\alpha=\beta-\alpha'=\beta+\underset{\text{emission}}{\beta_3}-\underset{\text{absorption}}{\beta_1}>0\tag{2-43}$$

Then (2-42) becomes

$$\frac{q^2\alpha'}{2m\beta}(\beta - \alpha') = \frac{q^2\alpha}{2m\beta}(\beta - \alpha) \tag{2-44}$$

In the following we define the α's to be

$$\left.\begin{aligned} \alpha &= \beta_e - \beta_a , & \beta_e > \beta_a \\ &= \beta + \beta_e - \beta_a , & \beta_e < \beta_a \end{aligned}\right\} \tag{2-45}$$

We can now calculate the contribution of the cluster of N 1-torons.

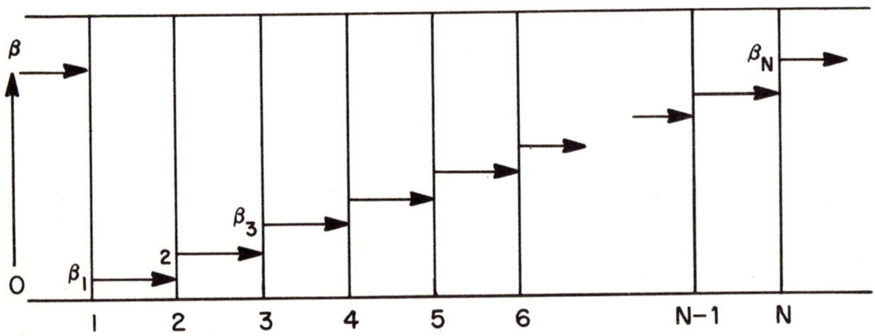

$$F_N = \frac{1}{(2\pi\hbar)^{3N}} \int dq_1 \cdots dq_N\, u(q_1) \cdots u(q_N) \int_0^{\beta} d\beta_N \int_0^{\beta_N} d\beta_{N-1} \cdots \int_0^{\beta_2} d\beta_1$$
$$\times\, G(q_1\,|\alpha_1|)\, G(q_2\,|\alpha_2|) \cdots G(q_N\,|\alpha_N|) \int dr_1\, dr_2 \cdots dr_N \tag{2-46}$$
$$\times\, \exp -\left[i\,\frac{q_1}{\hbar}\cdot(r_1 - r_2) + \frac{q_2}{\hbar}\cdot(r_2 - r_3) + \cdots + \frac{q_N}{\hbar}\cdot(r_N - r_1)\right]$$

where

$$G(q\,|\alpha|) = \frac{1}{(2\pi\hbar)^3}\left(\frac{2\pi m}{\beta}\right)^{3/2} e^{-\frac{q^2}{2m}\alpha(\beta-\alpha)} \tag{2-47}$$

$$\alpha_1 = \beta + \beta_1 - \beta_N, \quad \alpha_2 = \beta_2 - \beta_1, \cdots, \quad \alpha_N = \beta_N - \beta_{N-1} \tag{2-48}$$

Rearranging the factors in the exponential

$$q_1 \cdot (r_1 - r_2) + q_2 \cdot (r_2 - r_3) + \cdots + q_N \cdot (r_N - r_1)$$

$$= (q_1 - q_N) \cdot r_1 + (q_2 - q_1) \cdot r_2 + \cdots + (q_N - q_{N-1}) \cdot r_N$$

We find that the integrations over $r_1 \; r_2 \cdots r_N$ contribute the delta functions

$$\frac{1}{V} \int e^{i\frac{(q-q')}{\hbar} \cdot r} \, dr = \delta(q, q').$$

$$F_N = \frac{V}{(2\pi\hbar)^{3N}} \int dq \;\; u(q)^N \int_0^{\beta} d\beta_N \int_0^{\beta_N} d\beta_{N-1} \cdots \int_0^{\beta_2} d\beta_1$$

$$G(q\,|\alpha_1|)\,G(q|\,\alpha_2\,|)\,G(q|\,\alpha_3\,|) \cdots G(q|\,\alpha_N|) \qquad (2\text{-}49)$$

Here we want to eliminate the restriction on the ranges of integration

$$0 \le \beta_1 \le \beta_2 \le \cdots \le \beta_N \le \beta$$

$$0 \le \beta_{p_1} \le \beta_{p_2} \le \cdots \le \beta_{p_N} \le \beta \qquad N! \text{ permutations}$$

Whenever the emission temperature becomes lower than the absorption temperature a new α is defined by (2-45). Then

$$F_N = \frac{V}{(2\pi\hbar)^3} \cdot \frac{1}{N!} \int dq[u(q)]^N \int_0 \cdots \int_0 d\beta_N \ldots d\beta_1 \, G(q\,|\alpha_1|) \ldots G(q\,|\alpha_N!)$$

$$= \frac{V}{(2\pi\hbar)^3} \frac{1}{N!} \int dq[u(q)]^N \int_0 \cdots \int_0 d\alpha_1 \ldots d\alpha_N$$

$$\frac{1}{\beta} \sum_{n=-\infty}^{\infty} \exp \frac{2\pi i n}{\beta} [m\beta - \Sigma\,\alpha]$$

$$G(q\,|\alpha_1|)\,G(q\,|\alpha_2\,|) \cdots G(q\,|\alpha_N|) \qquad (2\text{-}50)$$

The factor

$$\frac{1}{\beta}\sum_{n=-\infty}^{\infty} \exp\frac{2\pi i n}{\beta}[m\beta - \Sigma\alpha]$$

is due to the fact that not all α's are independent but they satisfy the condition

$$\alpha_j = n_j\,\beta + \beta_{je} - \beta_{ja} \qquad n_j = 0 \text{ or } 1$$

$$\Sigma\alpha_j = m\,\beta, \qquad \text{m: The number of times of temperature inversion.}$$

In (2-50) the integrations over α's are independent and we obtain

$$F_N = \frac{V}{(2\pi\hbar)^3}\cdot\frac{1}{N!}\int dq \sum_n [\lambda_n(q)u(q)]^N \tag{2-51}$$

where

$$\lambda_n(q) = \frac{1}{\beta}\int_0^{\beta} G(q|\alpha|)\,e^{\frac{2\pi i n\alpha}{\beta}}\,d\alpha \tag{2-52}$$

is the finite Fourier transform of $G(q|\alpha|)$.

If we take into account that fact that there are $(N-1)!/2$ ways of constructing a ring of N 1-torons, the grand partition function is obtained by

$$\frac{pV}{kT} = \log\ Z_G^{(0)} + \frac{1}{2}\frac{V}{(2\pi\hbar)^3}\sum_n\int\sum_{N=2}^{\infty}\frac{(-1)^N}{N}[u(q)\lambda_n(q)]^N\,dq$$

$$= \log Z_G^{(0)} + \frac{1}{2}\frac{V}{(2\pi\hbar)^3}\int_0^{\infty} 4\pi q^2\,dq \sum_n \{u(q)\lambda_n(q) -$$

$$- \log[1 + u(q)\lambda_n(q)]\} \tag{2-53}$$

We now show that the result obtained for the Boltzmann gas is also valid for the ring contribution to the equation of Fermions and Bosons if the values of λ_n are the Fourier transform of an appropriate $G(q, |\alpha|)$.

In order to find $G(q, |\alpha|)$, we shall analyse the structure of an N-toron:

$$\int K_0(r\beta; r^{(M)}\beta_M) V(r^{(M)}) K_0 V K_0 \cdots K_0(r^{(n+1)}\beta_{n+1}; r^{(n)}\beta_n) \underline{V(r^{(n)})}$$

$$\underline{K_0(r^{(n)}\beta_n; r^{(n-1)}\beta_{n-1}) V(r^{(n-1)})} K_0 \cdots V(r^{(1)})$$

$$K_0(r^{(1)}\beta_1; r\,0)\, dr\, d\beta$$

The underlined factors are further analyzed in detail

$$V(R^{(n)} - r_j^{(n)}) K_0(r_N^{(n)}\beta_n; r_{N-1}^{(n-1)}\beta_{n-1}) K_0(r_{N-1}^{(n)}\beta_n; r_{N-2}^{(n-1)}\beta_{n-1})$$

$$\cdots K_0(r_j^{(n)}\beta_n; r_{j-1}^{(n-1)}\beta_{n-1}) K_0(r_{j-1}^{(n)}\beta_n; r_{j-2}^{(n-1)}\beta_{n-1})$$

$$\cdots K_0(r_2^{(n)}\beta_n; r_1^{(n-1)}\beta_{n-1}) K_0(r_1^{(n)}\beta_n; r_N^{(n-1)}\beta_{n-1})$$

$$\cdot\ V(r_1^{(n-1)} - r^{(n-1)}) \qquad (2\text{-}54)$$

This represents a toron of N-particles, a cycle of N propagators which is obtained by expanding the determinant $K_0(r^{(n)}\beta_n; r^{(n-1)}\beta_{n-1})$. We should notice that the temperature at which the propagators are created is not zero but β_{n-1}, and the temperature at which the propagators are annihilated is not β but β_n. However, because of the following reasons β_n and β_{n-1} will be reduced to β and 0 respectively.

For instance, the N-th particle is not interacting with other particles except the n - 1 st and the 1-st particles in the K(n, n-1) propagator, there exist factors like

$$K_0(r_N\beta; r_N^{(M)}\beta_M) K_0(r_N^{(M)}\beta_M) r_N^{(M-1)}\beta_{(n-1)}) \cdots K_0(r_N^{(n+1)}\beta_{n+1};$$

$$r_N\ \ \beta_n)$$

and

$$K_0(r_N^{(n-1)}\beta_{n-1}; r_N^{(n-2)}\beta_{n-2}) \cdots K_0(r_N^{(1)}\beta_1; r_N 0)$$

So that by the convolution type integrals over $r_N^{(M)} \cdots r_N^{(n-1)}$, $r_N^{(n)} r^{(n-1)} \cdots r^{(1)}$ the factors $K_0(r_N^{(n)}\beta_n; r_N^{(n-1)}\beta_{n-1})$ and $K_0(r_N^{(n)}\beta_n; r_{N-1}\ \ \beta_{n-1})$ become respectively

$$K_0(r_N\beta; r_{N-1}^{(n-1)}\beta_{n-1}),\quad K_0(r_1^{(n)}\beta_n; r_N\, 0)$$

By the same reason the other K_0's in (2-54) are also reduced and we obtain the following form

$$V(R^{(n)} - r_j^{(n)})\, K_0(r_1\beta;\, r_N 0)\; K_0(r_N\beta;\, r_{N-1}0) \cdots K_0(r_{j+1}\beta;\, r_j 0)$$

$$\times\, K_0(r_j\,\beta;\, r_j^{(n)}\beta_n)\, K_0(r_j\quad \beta_n;\, r_{j-1}0)\, K_0(r_{j-1}\beta;\, r_{j-2}0)\cdot$$

$$\cdots K_0(r_2\beta;\, r_1^{(n-1)}\beta_{n-1})\, K_0(r_1^{(n-1)}\beta_{n-1};\, r_1 0)\, V(r_1^{(n-1)} - r^{(n-1)}). \tag{2-55}$$

It is convenient to represent (2-55) pictorially.

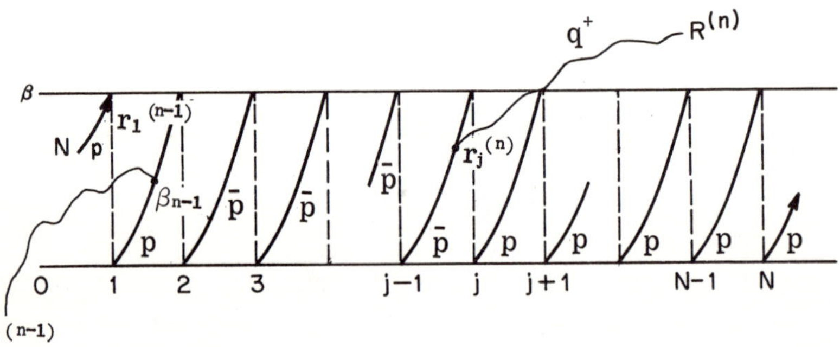

We now calculate the contribution of this toron to the grand partition function.

(Propagator 1→2), $\quad K_0(2;\,1) = \dfrac{1}{V}\displaystyle\sum_p e^{\,i\frac{p}{\hbar}\cdot(r_2-r_1) - \frac{p^2}{2m}(\beta_2-\beta_1)}$

Then (2-55) gives

$$\frac{1}{V^{N+2}}\sum_{p_1}\cdots\sum_{p_{N+2}}\int e^{\,i\frac{p_1}{\hbar}\cdot(r_1^{(n-1)}-r_1) - \frac{p_1^2}{2m}\beta_{n-1}}\, e^{\,i\frac{\bar{p}_1}{\hbar}(r_2-r_1^{(n-1)}) - \frac{\bar{p}_1^2}{2m}(\beta-\beta_{n-1})}$$

$$\times V(r - r_1^{(n-1)})\, V(r_j^{(n)} - R^{(n)})$$

$$\times\, e^{\,i\frac{p_2}{\hbar}\cdot(r_3-r_2) + \frac{p_3}{\hbar}\cdot(r_4-r_3) + \cdots + \frac{p_{j-2}}{\hbar}(r_{j-1}-r_{j-2}) - \frac{\beta}{2m}(p_2^2 + p_3^2 + \cdots + p_{j-2}^2)}$$

$$\times\, e^{\,i\frac{p_{j-1}}{\hbar}(r_j^{(n)}-r_{j-1}) - \frac{p_{j-1}^2}{2m}\beta_n}\quad e^{\,i\frac{\bar{p}_{j-2}}{\hbar}(r_j-r_j^{(n)}) - \frac{\bar{p}_{j-1}^2}{2m}(\beta-\beta_n)}$$

$$\times\, e^{\,i\frac{p_j}{\hbar}\cdot(r_{j+1}-r_j) + \cdots + \frac{p_N}{\hbar}\cdot(r_1-r_N)\; - \frac{\beta}{2m}(p_j^2 + \cdots + p_N^2)}\, dr_1\, dr_2 \cdots dr_N dr_j^{(n)} \tag{2-56}$$

$$\frac{1}{V}\int dr_1 \quad \rightarrow \quad \delta_{p_1\, p_N},\ p_1 = p_N \qquad \Big| \qquad \frac{1}{V}\int dr_j \quad \rightarrow \bar{p}_{j-1} = p_j$$

$$\frac{1}{V}\int dr_2 \quad \rightarrow \quad \bar{p}_1 = p_2 \qquad \Big| \qquad \frac{1}{V}\int dr_{j+1} \rightarrow p_j = p_{j+1}$$

$$\cdots \qquad \Big| \qquad \cdots$$

$$\frac{1}{V}\int dr_{j-1} \rightarrow \quad p_{j-2} = p_{j-1} \qquad \Big| \qquad \frac{1}{V}\int dr_N \rightarrow p_{N-1} = p_N$$

$$\bar{p}_1 = p_2 = p_3 = \cdots = p_{j-1} = \bar{p}, \quad \bar{p}_{j-1} = p_j = p_{j+1} = \cdots = p_N = p_1 = p$$

(2-56) now becomes

$$\left(\frac{1}{V}\right)^2 \sum_p \sum_{\bar{p}} \int e^{\,i\frac{(p-\bar{p})\cdot r_1^{(n-1)}}{\hbar} - \frac{p^2}{2m}\beta_{n-1} - \frac{\bar{p}^2}{2m}(\beta-\beta_{n-1})}\, e^{-\frac{(j-3)}{2m}\bar{p}^2\beta}$$

$$e^{-i\frac{p-\bar{p}}{\hbar}\cdot r_j^{(n)} - \frac{\bar{p}^2}{2m}\beta_n - \frac{p^2}{2m}(N-j+1)\beta}\ V(r_j^{(n)} - R^{(n)}) \tag{2-57}$$

Since

$$\left[u(q) = \int e^{-q\cdot r\hbar}\ V(r)\,dr,\ \ V(r) = \frac{1}{V}\sum_q e^{-iq\cdot r/\hbar}u(q)\right]$$

$$= \left(\frac{1}{V}\right)^3 \sum_p \sum_{\bar{p}} \sum_q \int dr_j^{(n)}\ e^{i\frac{p-\bar{p}}{\hbar}\cdot(r_1^{(n-1)} - r_j^{(n)})}\ e^{-\frac{p^2}{2m}[(N-j+2)\beta-\beta_n+\beta_{n-1}]}$$

$$e^{-\frac{\bar{p}^2}{2m}[(j-2)\beta+\beta_n-\beta_{n-1}]}\ e^{iq\cdot(r_j^{(n)}-R)/\hbar}\ u(q)$$

Since

$$\left[\frac{1}{V}\int dr_j^{(n)}\ e^{-\frac{(p-\bar{p}+q)}{\hbar}\cdot r_j^{(n)}} = \delta_{(\bar{p},\,p+q)}\right]$$

$$= \left(\frac{1}{V}\right)^2 \sum_p \sum_q e^{-i\frac{q}{\hbar}\cdot(r_1^{(n-1)}-R)}\ u(q)\ e^{-\frac{p^2}{2m}[(N-j+2)\beta-(\beta_n-\beta_{n-1})]}$$

$$e^{-\frac{(p+q)^2}{2m}[(j-2)\beta+\beta_n-\beta_{n-1}]} \tag{2-58}$$

$r_j^{(n)}$ can be any one of N coordinates (2-58) must be summed over $j[2 \sim N+1]$. Furthermore N can be 1, 2, $\cdots$, ∞. The contribution of torons can now be written as

$$\frac{1}{V} \sum_q e^{-i\frac{q}{\hbar}\cdot(r_1^{(n-1)}-R)} G(q, \alpha)$$

where

$$G(q, \beta_n - \beta_{n-1}) = \frac{1}{V} \sum_p \sum_{N=1}^{\infty} (-1)^{N+1} z^N \sum_{j=0}^{N-1} e^{-\frac{p^2}{2m}[(N-j)\beta-(\beta_n-\beta_{n-1})]}$$

$$e^{-\frac{(p+q)^2}{2m}[j\beta+(\beta_n-\beta_{n-1})]} \tag{2-59}$$

λ_n is now calculated by the formula

$$\lambda_n = \int_0^\beta G(q, \alpha)\, e^{2\pi i n\alpha/\beta} d\alpha \tag{2-60}$$

The Boltzmann form for $G(q, \alpha)$ yields

$$\lambda_n = \frac{z}{\hbar^3}\left(\frac{m}{\beta}\right)^{3/2} \int_0^\beta \exp\left[2\pi i n\alpha/\beta - \frac{q^2\alpha(\beta-\alpha)}{2m\beta}\right] d\alpha$$

$$q \to 0$$

as we will see later small q values have the biggest contribution.

$$\cong \frac{z}{\hbar^3}\left(\frac{m}{\beta}\right)^{3/2} \int_0^\beta \exp^{\frac{2\pi i n\alpha}{\beta}} \left[1 - \frac{q^2\alpha(\beta-\alpha)}{2m\beta}\right]$$

$$= \frac{z}{\hbar^3}\left(\frac{m}{\beta}\right)^{3/2} \beta\left[\delta_{n0}\left(1 - \frac{q^2\beta}{12m}\right. + \frac{q^2\beta}{4\pi^2 n^2 m}\right] \tag{2-61}$$

For $n \neq 0$,

$$-\int_0^\beta e^{\frac{2\pi i n\alpha}{\beta}} \frac{q^2\alpha(\beta-\alpha)}{2m\beta} d\alpha$$

$$= -\frac{1}{2\pi i n} e^{\frac{2\pi i n\alpha}{\beta}} \frac{q^2\alpha(\beta-\alpha)}{2m}\Bigg|_0^\beta + \frac{q^2}{2\pi i n}\int_0^\beta e^{\frac{2\pi i n\alpha}{\beta}} \frac{(\beta-2\alpha)}{2m} d\alpha$$

$$= -\frac{q^2\beta}{8\pi^2 n^2 m} e^{\frac{2\pi i n\alpha}{\beta}} (\beta - 2\alpha)\Big|_0^\beta - \frac{q^2\beta}{4\pi^2 n^2 m} \int_0^\beta e^{\frac{2\pi i n\alpha}{\beta}} d\alpha \Big\|_0$$

$$= \frac{q^2\beta^2}{4\pi^2 n^2 m}$$

For Fermion case: From (2-59) and (2-60)

$$\lambda_n = \frac{1}{V}\sum_p \sum_{N=1}^{\infty} (-1)^{N+1} z^N \sum_{j=0}^{N-1} \int_0^\beta e^{-\frac{p^2}{2m}[N\beta-(j\beta+\alpha)] - \frac{(p+q)^2}{2m}(j\beta+\alpha)} e^{\frac{2\pi i n}{\beta}[j\beta+\alpha]} d\alpha$$

$$= \frac{1}{V}\sum_p \sum_{N=1}^{\infty} (-1)^{N+1} z^N \int_0^{N\beta} e^{-\frac{p^2}{2m}(N\beta-\alpha) - \frac{(p+q)^2}{2m}\alpha} e^{\frac{2\pi i n}{\beta}\alpha} d\alpha$$

$$= \frac{\beta}{V}\sum_p \sum_{N=1}^{\infty} (-1)^{N+1} z^N N \int_0^1 e^{-\frac{p^2}{2m}N\beta(1-\lambda) - \frac{(p+q)^2}{2m}N\beta\lambda + \frac{2\pi i n}{\beta}N\beta\lambda} d\lambda$$

$$= \frac{\beta}{V}\sum_p \sum_{N=1}^{\infty} (-1)^{N+1} z^N N \frac{1}{(2\pi)^2} \int_0^1 \iint_{-\infty}^{\infty} \frac{e^{iN\beta(1-\lambda)s_1}}{(i s_1 + \frac{p^2}{2m})}$$

$$\times \frac{e^{iN\beta\lambda s_2}}{i s_2 + \frac{(p+q)^2}{2m}} e^{\frac{2\pi i n}{\beta}N\beta\lambda} d\lambda\, ds_1\, ds_2$$

$$\frac{1}{2\pi}\int_{-\infty}^{\infty} \frac{e^{ias}}{is + (p^2/2m)} ds = c^{-a\frac{p^2}{2m}} \quad a > 0 \tag{2-62}$$

Now integration over λ contributes a delta function

$$\frac{N\beta}{2\pi}\int_0^1 e^{iN\beta s\lambda} d\lambda = \delta(s).\ e^{\frac{1}{2}iN\beta S} \quad (\text{for } \beta \to \infty)$$

so that

$$-s_1 + s_2 + \frac{2\pi n}{\beta} = 0, \quad s_2 = s_1 - \frac{2\pi n}{\beta}$$

$$\lambda_n = \frac{1}{V} \sum_p \sum_{N=1}^{\infty} (-1)^{N+1} z^N \int_{-\infty}^{\infty} ds \frac{e^{iN\beta s}}{(is + \frac{p^2}{2m})\left(\frac{(p+q)^2}{2m} + i[s - \frac{2\pi n}{\beta}]\right)}$$

$$= \frac{1}{V} \sum_p \int_{-\infty}^{\infty} ds \frac{1}{\left(\frac{p^2}{2m} + is\right)} \times \frac{1}{\frac{(p+q)^2}{2m} + i\left[s - \frac{2\pi n}{\beta}\right]} \times \frac{z e^{i\beta s}}{1 + z e^{i\beta s}} \tag{2-63}$$

$$= \frac{1}{(2\pi\hbar)^3} \int dp \int_{-\infty}^{\infty} ds \frac{1}{\left(\frac{p^2}{2m} + is\right)} \times \frac{1}{\frac{(p+q)^2}{2m} + i\left[s - \frac{2\pi n}{\beta}\right]} \frac{z e^{i\beta s}}{1 + z e^{i\beta s}}$$

Evaluation of the integral

$$I_0 = \lim_{\beta\to\infty} \int_{-\infty}^{\infty} ds \int dp \frac{1}{\left(\frac{p^2}{2m} + is\right)\left[\frac{(p+q)^2}{2m} + i(s - kt)\right]} \times \frac{z e^{i\beta s}}{1 + z e^{i\beta s}} \qquad \left[\begin{array}{l} e^{i\beta s} \to 0 \\ s \to i\infty \end{array}\right.$$

$$= \lim_{\beta\to\infty} (-2\pi i) \sum_\xi \int dp \frac{1}{\left[\frac{p^2 - p_0^2}{2m} + i\frac{(2\xi+1)\pi}{\beta}\right]} \times \frac{1}{\left[\frac{(p+q)^2}{2m} - \frac{p_0^2}{2m} + i\frac{(2\xi+1)\pi}{\beta} - ikt\right]} \times \frac{1}{i\beta}$$

Poles are

$$z e^{is\beta} = -1 \qquad is\beta + \frac{p_0^2}{2m}\beta = (2\xi + 1)\, i\pi$$

$$e^{is\beta + \frac{p_0^2}{2m}\beta} = -1, \qquad s = i\frac{p_0^2}{2m} + \frac{(2\xi+1)\pi}{\beta}$$

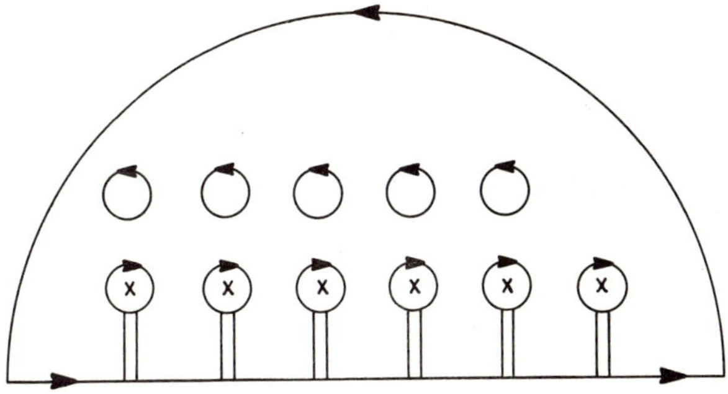

In the limit $\beta \to \infty$ (low temperature)

$$\frac{2\xi + 1}{\beta}\,\pi = x;\ 2\xi + 1 = \frac{1}{\xi}\frac{\beta}{\pi}x;\ \xi + \frac{1}{2} = \frac{\beta}{2\pi\xi}\,x,$$

$$\sum = \left(\frac{\beta}{2\pi i}\right)\,dx$$

$$I_0 = \int dp \int_{-\infty}^{\infty} dx\,\frac{1}{\left[x - i\left(\frac{p^2 - p_0^2}{2m}\right)\right]}\,\frac{1}{\left[x - i\,\frac{[(p+q)^2 - p_0^2]}{2m} - kt\right]} \tag{2-64}$$

The x integral gives a non-zero contribution only if either

$$p^2 > p_0^2, \qquad (p+q)^2 < p_0^2$$

or

$$p^2 < p_0^2, \qquad (p+q)^2 > p_0^2$$

This is clearly the effect of the exclusion principle at absolute zero, to our approximation. We can write

$$I_0 = 2\pi \int dp\,\theta(p)\left[\frac{(p+q)^2}{2m} - \frac{p^2}{2m} - ikt\right]^{-1} \tag{2-65}$$

where

$$\theta(p) = \begin{cases} +1 \text{ if } (p+q)^2 > p_0^2,\ p^2 < p_0^2 \\ -1 \text{ if } (p+q)^2 < p_0^2,\ p^2 > p_0^2 \\ \ \ 0 \text{ otherwise} \end{cases}$$

For $q \ll p_0$, this integral has been calculated by Gell-Mann and Brueckner. The result is, as $g \to 0$ and $\beta \to \infty$

$$8\pi^2 p_0 m\{1 - u \tan^{-1} u^{-1}\}$$

where

$$u \equiv kt\, m/qp_0$$

Using this value

$$\lambda_n \sim \frac{2m}{\hbar^3}\frac{1}{(2\pi)^{1/2}} p_0[1 - v\tan^{-1} v], \quad v = 2\pi n m/\beta q p_0 \qquad (2\text{-}66)$$

The fugacity or the chemical potential $p_0^2/2m$ is obtained in terms of the density through

$$\rho = \frac{N}{V} = \frac{z}{V}\frac{\partial \log Z_G}{\partial z}$$

$$= \frac{z}{V}\frac{\partial \log Z_G^{(0)}}{\partial z} + \frac{1}{2}(2\pi\hbar)^{-3}\int 4\pi q^2\, dq \sum_j \frac{u^2(q)\lambda_j}{1+\lambda_j u(q)} \cdot \frac{\partial \lambda_j}{\partial \log Z}$$

$$(2\text{-}67)$$

$$\frac{\partial}{\partial \log z}\{u(q)\lambda_j - \log[1 + u(q)\lambda_j]\}$$

$$= u(q)\frac{\partial \lambda_j}{\partial \log z} - \frac{u(q)}{1 + u(q)\lambda_j}\frac{\partial \lambda_j}{\partial \log z}$$

$$= \frac{u^2(q)\lambda_j}{1 + u(q)\lambda_j} \cdot \frac{\partial \lambda_j}{\partial \log z}$$

The internal energy is given by

$$E = -\frac{\partial \log Z_G}{\partial \beta}$$

$$= -\frac{\partial \log Z_G^{(0)}}{\partial \beta} - \frac{1}{2}V(2\pi\hbar)^{-3}\int 4\pi q^2\, dq \sum_j \frac{u^2(q)\lambda_j}{1+u(q)\lambda_j}\frac{\partial \lambda_j}{\partial \beta}$$

$$(2\text{-}68)$$

The ground-state energy is the limit of this quantity as $\beta \to \infty$.

One important low-order diagram which was omitted from our sum of ring integrals is the self (or exchange) interaction of a toron. This is a process in which a quantum of momentum is emitted from an N toron at some point at reciprocal temperature β' and absorbed at one of the reciprocal temperatures $\beta' + \beta$, $\beta' + 2\beta$, $\cdots$, $\beta' + (N-1)\beta$. Since such a process cannot occur in a Boltzmann system the contribution of the corresponding integrals need only be added to $\log Z^G$ in the fermion or boson case.

Note of the editor:

Montroll has given in his later work (Colorado Summer School 1960, page 290), a simple rule to write down integrals such as (2-35). We will illustrate this rule for the case mentioned in the text: the term in (2-35) that uses the underlined potentials. The picture used with it on page 82 consists essentially of 9 regions. Four of these regions are related to propagators (1, 2, 3, 4, 5 and 6) and three regions are related to interactions (7, 8 and 9).

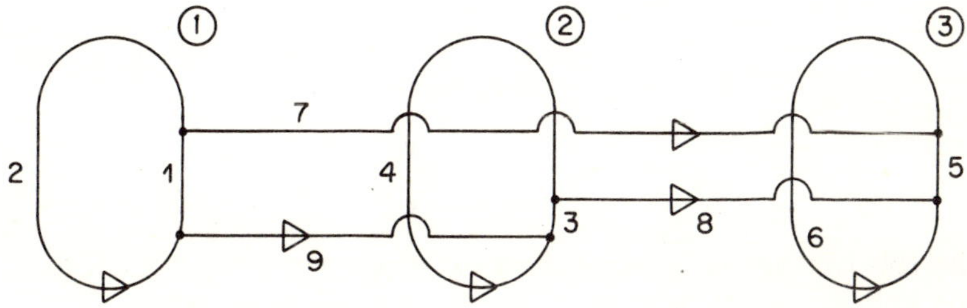

The propagator depends only on the relative distance Δr and on the length l in β-space

$$K(r_2\beta_2 ; r_1\beta_1) \equiv K(1, \overrightarrow{\Delta r}) =$$

$$= \frac{1}{(2\pi\hbar)^3} \int e^{-lp^2/2m} \, e^{i\vec{\Delta r}\cdot\vec{p}/\hbar} \, d^3p.$$

we have (notice that with r' goes β_1 with r" goes β_2 and with r‴ goes β_3).

region	1	Δr
1	$\beta_1 - \beta_3$	$r_1' - r_1'''$
2	$\beta - (\beta_1 - \beta_3)$	$r_1''' - r_1'$
3	$\beta_2 - \beta_3$	$r_2'' - r_2'''$
4	$\beta - (\beta_2 - \beta_3)$	$r_2''' - r_2''$
5	$\beta_1 - \beta_2$	$r_3' - r_3''$
6	$\beta - (\beta_1 - \beta_2)$	$r_3'' - r_3'$

There are six space integrations involved, each assoicated with a black dot. There are nine momentum integrations involved since each region is represented by a Fourier integral. For the region 1-6 they are given above for 7-9 we have:

$$\phi(\Delta r) = \int u(q)\; e^{iq\Delta r/\hbar}\; d^3q$$

region	Δr
7	$r_3' - r_1'$
8	$r_3'' - r_2''$
9	$r_2''' - r_1'''$

The total integral is actually a multiply connected convolu tion and since all factors in the integrand are written as fourier transforms of the relative distance we can expect certain simplifications upon integrating over the space variables. Before we integrate let us make the rule that each Fourier variable carries the label of its region: In region 1 we use p_1; in 2 use p_2; in 7 use q_7 etc.

If one now performs the six space integrations it is easy to see (by writing the exponentials that carry the same r together) that one obtains δ-function relations in each point

$$r_1''' \rightarrow (2\pi\hbar)^3\; \delta(p_2 - q_9 - p_1)$$
$$r_1' \rightarrow (2\pi\hbar)^3\; \delta(p_1 - q_7 - p_2)$$
$$r_2''' \rightarrow (2\pi\hbar)^3\; \delta(p_4 + q_9 - p_3)$$

$$r_2'' \rightarrow (2\pi\hbar)^3 \ \delta(p_3 - q_8 - p_4)$$
$$r_3'' \rightarrow (2\pi\hbar)^3 \ \delta(p_6 + q_8 - p_5)$$
$$r_3' \rightarrow (2\pi\hbar)^3 \ \delta(p_5 + q_7 - p_6)$$

An error has been made since the last integration was superfluous. The five previous δ-functions imply the sixth, hence the last exponential was $e^0 = 1$ and the last integration contributes V, the volume of the container.

The result is an integral over nine momentum variables containing 5 (independent) δ-functions so that there are actually four momentum integrals for which we (arbitrarily) take p_1, p_2, p_3 and q_9. The temperature exponential is (reading from the table)

$$(\beta_1 - \beta_3)\, p_1^2 + (\beta - \beta_1 + \beta_3)\, p_2^2 + (\beta_2 - \beta_3)\, p_3^2 + (\beta - \beta_2 + \beta_3)\, p_4^2$$
$$+ (\beta_1 - \beta_2)\, p_5^2 + (\beta - \beta_1 + \beta_2)\, p_6^2 \equiv f(p_1 p_2 p_3 p_4 p_5 p_6 ; \beta, \beta_1 \beta_2 \beta_3).$$

The complete cluster integral for this particular term is (remember that the kernels K are only non-zero for positive temperature interval)

$$\frac{V}{(2\pi\hbar)^3} \int_0^\beta d\beta_1 \int_{\beta_1}^\beta d\beta_2 \int_{\beta_2}^\beta d\beta_3 \iiiint dp_1\, dp_2\, dp_3\, dq\ u(-q)u(q)u(q)$$
$$e^{-f/2m}.$$

In this integral p_4 p_5 p_6 are determined by the five momentum relations; they imply also that $q_7 = -q$, $q_8 = q$ and $q_9 - q$.

2.5 EQUATION OF STATE OF THE ELECTRON GAS

Equation (2-53) can be applied to the theory of the electron gas. In the classical limit ($\hbar \rightarrow 0$) it yields the Debye-Hückel theory while at low temperatures ($\beta \rightarrow \infty$) it leads to the Gell-Mann-Brueckner equation for the correlation energy.

Although both the Debye-Hückel and Gell-Mann-Brueckner

formulas are $e^2 \to 0$ limits the Debye-Hückel theory is also a low density limit while the Gell-Mann-Brueckner theory is a high density limit. This apparent paradox is resolved by a simple dimensional argument.

The thermodynamic properties of an electron gas are functions of the five independent parameters; the electron mass m, the charge e, the density ρ, k and $\beta = 1/kT$. The dimensionless energy, say, the ratio of the internal energy to the Fermi energy must be a function of dimensionless combinations of these parameters. The constant h does not enter into the combination in the classical limit and the required dimensionless variable is $e^2 \beta\rho^{1/3}$ while as $\beta \to \infty$ (ground-state properties) the appropriate variable is $me^2/\hbar\rho^{1/3}$. Hence in the classical limit any expansion in powers of e would also be one in powers of density (DH) and therefore a low density expansion, while in the low temperature limit an expansion in powers of e would also be one in inverse powers of the density (GM - B) and therefore a high density expansion.

The only dimensionless parameter of the four variables β, m, $\hbar$, and ρ is $\beta\hbar^2\rho^{2/3}/m$. Hence a general expression which gives a thermodynamic property of an electron gas at any temperature would contain the electronic charge in the combination $me^2\rho^{-1/3}\hbar^{-2} f(\beta\hbar^2\rho^{2/3}m^{-1})$ where $f(\chi)$ is an arbitrary function such that

$$f(\chi) \quad \begin{cases} 1 \text{ as } \beta \to \infty \\ \chi \text{ as } \beta \to 0 \end{cases}$$

The Fourier transform of the Coulomb potential $V(r) = e^2/r$ is obtained in the following way:

$$u(q) = \frac{e^2}{(2\pi)^{3/2}} \int \frac{e^{-\alpha r}}{r} e^{-q\cdot r\hbar^{-1}} \overrightarrow{dr}$$

$$= \frac{e^2\, 2\pi}{(2\pi)^{3/2}} \int_0^\infty \frac{e^{-\alpha r}}{r} r^2\, dr \int_0^\pi d\theta\, e^{iqr\hbar^{-1}\cos\theta} \sin\theta\, d\theta$$

$$= \frac{e^2}{(2\pi)^{1/2}} \frac{\hbar}{q} \frac{1}{i} \int_0^\infty \left[e^{-\alpha r + iqr\hbar^{-1}} - e^{-\alpha r - iqr\hbar^{-1}}\right] dr$$

$$= \frac{e^2}{(2\pi)^{1/2}} \frac{\hbar}{q} \frac{1}{i} \left[\frac{1}{\alpha - iq\hbar^{-1}} - \frac{1}{\alpha + iq\hbar^{-1}}\right]$$

$$= \frac{2e^2}{(2\pi)^{1/2}} \cdot \frac{1}{\alpha^2 + q^2\hbar^{-2}} \rightarrow \left(\frac{2}{\pi}\right)^{1/2} \hbar^2 \frac{e^2}{q^2}, \text{ as } \alpha \rightarrow 0 \quad (2\text{-}69)$$

We now derive the DH equation of state. It was shown (2-61) that when the momentum transfer q per collision is small in a Boltzmann electron gas $\lambda_0 \sim$ constant while $\lambda_k \sim q^2$ for $\neq 0$. Hence with the inverse square form of u(q), (2-69), for the electron gas we see that

$$(2\text{-}51) \rightarrow F_N = \frac{V}{(2\pi\hbar)^3} \frac{1}{N!} \int dq \sum_n [\lambda_n(q) u(q)]^N \quad (2\text{-}70)$$

the small q behavior of the integrand of (2-70) with n = 0 causes a divergence of the individual ring integrals. On the other hand we see from (2-53)

$$(2\text{-}53) \rightarrow \frac{pV}{kT} = \log Z_G^{(0)} + \frac{1}{2} \frac{V}{(2\pi\hbar)^3} \int_0^\infty 4\pi q^2 \, dq \quad (2\text{-}71)$$

$$\sum_n \{ u(q)\lambda_n(q) - \log[1 + u(q)\lambda_n(q)] \}$$

that the sum of all the ring integrals does not diverge as a result of low momentum transfer processes. However the divergence of the individual ring integrals implies that the main contribution to thermodynamic properties of an electron gas comes from the small q range of the integrand of (5.14). In view of (2-61)

$$(2\text{-}61) \rightarrow \lambda_n = \frac{z}{\hbar^3} \left(\frac{m}{\beta}\right)^{3/2} \beta \left[\delta_{n,0} + \frac{q^2 \beta}{4\pi^2 n^2 m}\right] \quad (2\text{-}72)$$

The main contribution to the n summation as $q \rightarrow 0$ comes from n = 0. Since as $q \rightarrow 0$

$$\lambda_0 - z\hbar^{-3} \left(\frac{m}{\beta}\right)^{3/2} \beta \quad (2\text{-}73)$$

and since $u(q) = (2/\pi)^{1/2} \hbar^2 e^2/q^2$ (2-71) yields

$$\frac{pV}{kT} = \frac{zV\hbar^{-3}}{(2\pi)^{3/2}(\rho/m)^{3/2}} + \frac{2\pi V\hbar^{-3}}{(2\pi)^3}\int_0^\infty \{sz - q^2 \log(1 + szq^{-2})\}\,dq \tag{2-74}$$

where

$$s = e^2 \left(\frac{2}{\pi}\right)^{1/2} \frac{\beta}{\hbar} \left(\frac{m}{\beta}\right)^{3/2}$$

In (2-74) the q integral is elementary

$$\int_0^\infty [sz - q^2 \log(1 + szq^{-2})]\,dq$$

$$= (sz) \lim_{R\to\infty} \int_0^R [1 - \frac{q^2}{sz} \log(1 + \frac{sz}{q^2})]\,dq$$

$$\frac{q^2}{sz} = t^2, \quad dq = (sz)^{1/2}\,dt$$

$$= (sz)^{3/2} \lim_{R\to\infty} \int_0^R [1 - t^2 \log(1 + \frac{1}{t^2})]\,dt$$

$$= \frac{\pi}{3}(sz)^{3/2}$$

$$\lim_{R\to\infty} \int_0^R [1 - t^2 \log(1 + \frac{1}{t^2})]\,dt$$

$$= \lim_{R\to\infty} \left[R - \frac{t^3}{3}\log(1 + \frac{1}{t^2})\Big|_0^R - \frac{2}{3}\int_0^R t^3 \frac{\frac{1}{t^3}}{1 + \frac{1}{t^2}}\,dt\right]$$

$$= \frac{2}{3}\lim_{R\to\infty} \left[R - \int_0^R \frac{t^2}{1 + t^2}\,dt\right.$$

$$= \frac{2}{3}\lim_{R\to\infty} \int_0^R \frac{dt}{1 + t^2}$$

$$= \frac{\pi}{3}$$

$$\frac{pV}{kT} = \frac{zV\hbar^{-3}}{(2\pi)^{3/2}(\beta/m)^{3/2}} + \frac{V\hbar^{-3}}{(2\pi)^2}\cdot\frac{\pi}{3}\left[ze^2\left(\frac{2}{\pi}\right)^{1/2}\frac{\beta}{\hbar}\left(\frac{m}{\beta}\right)^{3/2}\right]^{3/2} = \log Z_G \tag{2-75}$$

$$\rho = \frac{z}{V} \cdot \frac{\partial \log Z}{\partial z} = \frac{z\hbar^{-3}}{(2\pi\beta/m)^{3/2}}$$

$$+ \frac{\pi\hbar^{-3}}{2(2\pi)^2}\left\{z e^2 \left(\frac{2}{\pi}\right)^{1/2}\frac{\beta}{\hbar}\left(\frac{m}{\beta}\right)^{3/2}\right\}^{3/2} \qquad (2\text{-}76)$$

$$\rho \to 0, \qquad z_0 = (2\pi\hbar^2\beta/m)^{3/2}\rho$$

$$z = (2\pi\hbar^2\beta/m)^{3/2}\rho\{1 - e^3\beta^{3/2}\rho^{1/2}\pi^{1/2} + 0(\rho)\}$$

$$\frac{p}{kT} = \rho\left\{1 - \frac{1}{3}\pi^{1/2} e^3 \beta^{3/2}\rho^{1/2}\right\} \qquad (2\text{-}77)$$

The ionic strength

$$\kappa^2 = \frac{4\pi}{VkT}\ \Sigma N_i Q_i^2$$

(N_i being the number of charges of the i-th species with charge Q_i.)

Neutrality condition $N_+ Q_+ = N_- e$

continuum of positive charge $\quad N_+ \to \infty,\ Q_+ \to 0$

$$Q_+ = \frac{N_-}{N_+} e$$

$$\Sigma N_i Q_i^2 = N_- e^2 + N_+\left(\frac{N_-}{N_+} e\right)^2 = N_- e^2\left(1 + \frac{N_-}{N_+}\right) \to N_- e^2 \qquad (2\text{-}78)$$

Hence

$$\kappa^2 = \frac{4\pi}{kT} \cdot \frac{N_-}{V} e^2 = 4\pi\beta\rho e^2 \qquad (2\text{-}79)$$

Since the ionic strength has the dimension of a reciprocal length, $1/\kappa$ is the characteristic length (the Debye length) which measures the mean range of an electron in the classical electron gas.

As $\beta \to \infty$ the quantum (statistical) effect becomes important and the effect of torons of all orders in the ring integral contribution should be taken into account. In this case

$$\log Z_G = \log Z_G^{(0)} + \frac{1}{2} V(2\pi\hbar)^{-3} \int_0^\infty 4\pi q^2 \; dq \sum_n \{ 2\lambda_n u(q) - \log[1 + 2u(q)\lambda_n] \}, \tag{2-80}$$

where

$$\lambda_n = 2m\hbar^{-3}(2\pi)^{-\frac{1}{2}} p_0 [1 - v \tan^{-1}(1/v)], \quad v = \frac{2\pi m n}{\beta q p_0},$$

As $\beta \to \infty$, v becomes a continuous variable with

$$\sum_n \sim \frac{\beta q p_0}{2\pi m} \; dv$$

$$\log Z_G = \log Z_G^{(0)} + V(2\pi\hbar)^{-3} \frac{\beta p_0{}^5}{m} \int_0^\infty Q^3 \; dQ \int_{-\infty}^\infty \left\{ \frac{4me^2}{\hbar\pi Q^2 p_0} R(v) - \log\left[1 + \frac{4me^2}{\pi\hbar p_0 Q^2} R(v)\right] \right\} dv, \tag{2-82}$$

$$q = p_0 Q, \quad 2\lambda(v)\, u(q) = (4me^2 p_0/\hbar q^2 \pi)\, R(v),$$
$$R(v) = 1 - v \tan^{-1}(1/v). \tag{2-83}$$

$$\frac{E}{N} = -\frac{1}{N}\frac{\partial \log Z_G}{\partial \beta} = [2\rho m(2\pi\hbar)^3]^{-1} [8\pi p_0^5/5 + 3\, I(2m\mu/\beta)^{5/2}] \tag{2-84}$$

$$\rho = \frac{1}{V} \cdot \frac{\partial \log Z}{\partial \mu} = [2m(2\pi\hbar)^3]^{-1} [16 m\pi p_0{}^3/3 + 5\beta(2m\mu/\beta)^{5/2} I/\mu] \tag{2-85}$$

$$\mu = \log z = \beta p_0{}^2/2m$$

$$I = \int_0^\infty Q^3 \; dQ \int_{-\infty}^\infty \{ \qquad \} \; dv$$

$$E_{corr} = (E/N) - (E^{(0)}/N), \qquad p_0 = p_F + \delta p_0, \quad p_F = \hbar(3\pi^2\rho)^{\frac{1}{3}}$$

$$\alpha = \left(\frac{4}{q\pi}\right)^{\frac{1}{3}}, \quad r_s = r_0/r_B \qquad \rho = \frac{1}{V}\frac{\partial \log Z_G}{\partial \mu} = \frac{1}{V}\frac{\partial \log Z_G{}^{(0)}}{\partial \mu^{(0)}}$$

$$\left(\frac{4}{3}\right) \pi r_0{}^3 \rho = 1. \qquad \delta p_0/p_F = -5I/8\pi$$

$$E_{corr} = -\frac{3}{8\pi m}\left(\frac{9\pi}{4}\right)^{\frac{2}{3}}\frac{\hbar^2}{r_0{}^2}\int_0^\infty Q^3\,dQ\int_{-\infty}^{\infty}\left\{\frac{4\,\alpha r_s}{\pi Q^2}R(v)\right.$$

$$\left. -\log\left[1+\frac{4\,\alpha r_s}{\pi Q^2}R(v)\right]\right\}dv \qquad (2\text{-}86)$$

REFERENCES

E. W. Montroll and J. C. Ward, Phys. Fluids **1**, 55 (1958).

M. Gell-Mann and K. A. Brueckner, Phys. Rev., **106**, 364 (1957).

Diagrammatical Method

Tohru Morita

LINKED CLUSTER EXPANSION FOR THE LOGARITHM OF THE GRAND PARTITION FUNCTION

In the theory of the many-body problem there appear many multifold integrals and diagrams are introduced in order to visualize the structure of these integrals. Typical ones are the diagrams for the cluster integrals in classical statistical mechanics and the Feynmann diagrams in quantum-electrodynamics. In the following, a diagram is associated with an integral and we shall mean by a diagram the integral corresponding to the diagram.

In the many-body problem appear various sums of many multi-fold integrals which have some topological structure if we express them by the aid of diagrams. The topological structure of a set of diagrams (expressing integrals and corresponding to a physical quantity) suggest us sometimes to take a partial summation or renormalization. This process defines a function by a sum of diagrams constituting of a part of the original diagrams and expresses the sum of diagrams in terms of it. In this way we get a sum of new diagrams for the physical quantity we are interested in, which have a more restrictive topological structure than the original ones. For that purpose, it is very convenient to formulate the theorems which produce the partial summation or renormalization to sums of diagrams. These theorems make the use of diagrams a powerful tool.

1. DEFINITION OF DIAGRAMS.

Let us start with the definition of a diagram, which represents a product of factors. A diagram consists of lines and vertices (end points of lines). Usually a vertex is a common end point of two or more lines. If this is so the diagram is a closed diagram. Each line is a function of its own variable and/or of the variables (coordinates) of the two vertices at which the line ends. Each vertex is a function of the local coordinate (or simply the number one).

DEFINITION 1. A diagram represents a product of factors (lines and vertices) with the specified values of the arguments (i. e., coordinates and variables), divided by the symmetry number of the diagram.

The symmetry number of a diagram is the number of indistinguishable ways in labeling all of the elements by the numbers from one to the total number of elements.

Some examples of diagrams defined by Definition 1 and the corresponding explicit expressions are given in Fig. 1-1. The numerical factors in front shows the inverse of the symmetry number of the respective diagrams. These diagrams represent integrands.

DEFINITION 2. An "integral diagram" is a diagram which is obtained from the diagram for the integrand by removing some or all of the coordinates and variables, represents an integral. This integral is taken over the coordinates and variables which are removed from the product of factors corresponding to the original diagram for the integrand and is divided by the symmetry number of the diagram. It should be noticed that the symmetry number changes if we remove the coordinates.

There will be several possible distinguishable diagrams for the integrand for one diagram defined by Definition 2,

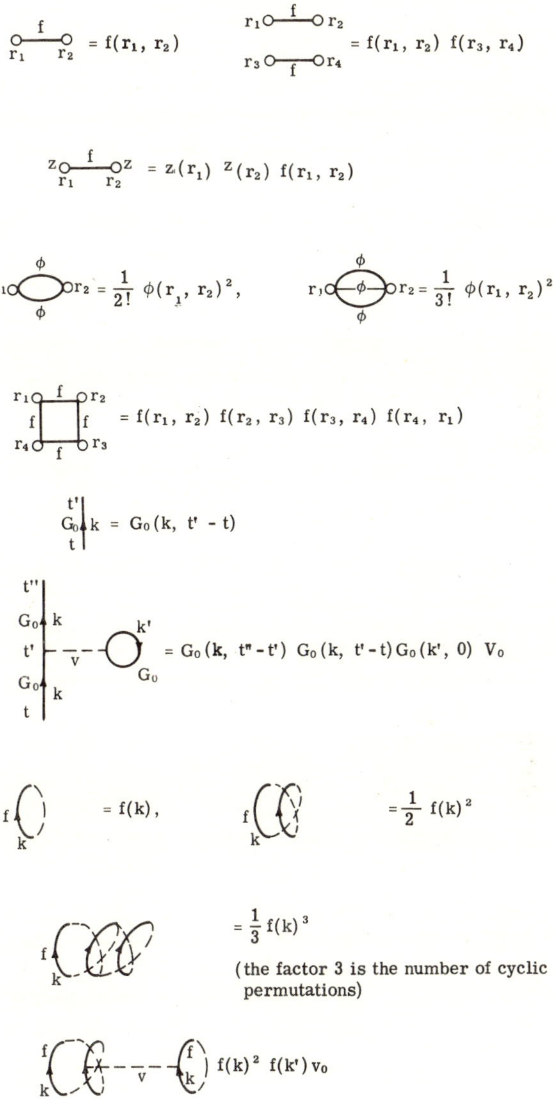

Fig. 1. Examples of diagrams defined by Def. 1.

even if we remove the set of coordinates and variables from the integrand (see the last example in figure 1-2). The number of distinguishable integrand diagrams is equal to the factorial of the number of variables (resp. coordinates) to be integrated, divided by the symmetry number of the diagram for the integral, multiplied by the symmetry number of the diagram for integrand. If there are more than one type of variable we have to take the product of factorials. Hence we can rewrite Def 2 as:

DEFINITION 2'. A diagram which is obtained from the diagram for the integrand by removing some or all of the coordinates and variables, represents an integral over the coordinates and variables which are removed. The integrand is the sum of the products of factors corresponding to all the possible different original diagrams for the integrand, which are obtained by interchanging the variables and coordinates over which integrations are performed; divided by the product of the symmetry number of the diagrams for the integrand and the factorial of the number of variables and coordinates of the same kind.

Examples are given in Fig. 1-2. The factor in front shows the inverse of the symmetry number of the diagrams.

2. BASIC THEOREMS.

The first step in the theory of many body problem is to get for any quantity expressions that have the correct order of magnitude (correct power of volume Ω) in the limit of $\Omega \to \infty$ while keeping the intensive quantities such as particle density, temperature and so on constant. In the diagrammatical approach, this first step is used to show the cancellation of disconnected diagrams and consequently that physical quantities can be expressed by a sum of connected diagrams. This situation is stressed, for instance, in the terminology like "linked cluster expansion" for the ground state energy of many body systems.

The quantity which is the most important in the equilibrium statistical mechanics is the free energy as a function

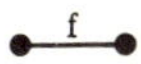

$$= \frac{1}{2}\int dr_1 \int dr_2\, f(r_1, r_2),$$

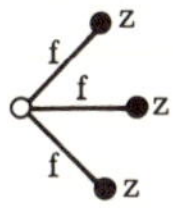

$$= \frac{1}{3!}\left[\int f(r', r)\,z(r)\,dr\right]^3$$

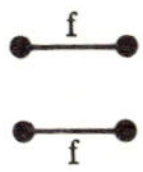

$$= \frac{1}{8}\left[\int dr_1 \int dr_2\, f(r_1,\ r_2)\right]^2$$

$$= \frac{1}{2!}\,\frac{1}{2!}\int dr_1 \int dr_2\, \phi(r_1,\ r_2)^2$$

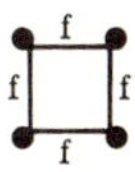

$$= \frac{1}{8}\int dr_1 \int dr_2 \int dr_3 \int dr_4\, f(r_1, r_2) f(r_2, r_3) f(r_3, r_4) f(r_4, r_1)$$

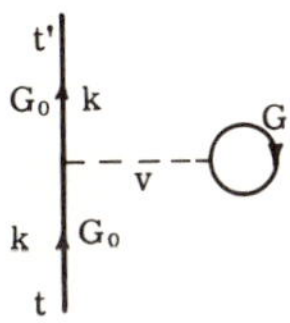

$$\sum_{k'} \int_t^{t'} dt''\, G_0(k, t' - t'')\, G_0(k, t'' - t)\;\; G_0(k',\ 0)\, v_0$$

$$= \frac{1}{2}\, f(k)^2$$

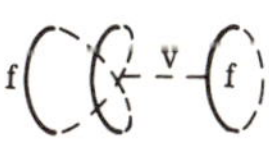

$$\Sigma_k\, \Sigma_{k'}\;\; f(r)^2 f(k')\, v_0 \;\; \int_0^\beta du$$

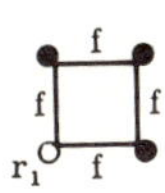

$$= \frac{1}{3!}\int dr_2 \int dr_3 \int dr_4 \left\{ \vphantom{\int} \right.$$

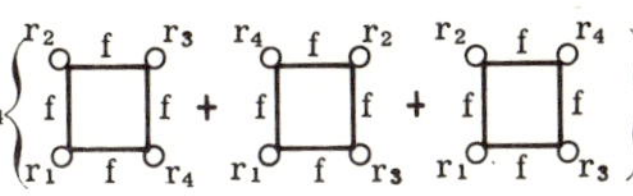

$$\left. \vphantom{\int} \right\}$$

(Symmetry number is 2)

Fig. 1-2. Examples of diagrams defined by Def. 2.

of the corresponding proper independent thermodynamic variables. There are different forms of the free energy, each having their own suitable set of independent variables. The knowledge of these as a function of their proper set of independent variables are known to be equivalent, so that we can use any of them according to our convenience. The one which is the most suitable for the diagrammatical approach is the logarithm of the grand partition function, $\ln \Xi$, which is equal to $p\Omega/kT$, p is pressure. The proper set of corresponding independent variables are the volume Ω, the temperature T or its inverse $\beta = 1/kT$ and the chemical potential μ or the fugacity $z = \exp \mu/kT$; where k is the Boltzmann constant. The variation of $\ln \Xi$ with respect to those of the independent variables is given by

$$\delta \ln \Xi = N \delta \ln z - E \delta\beta + \frac{p}{kT} \delta\Omega. \qquad (2.1)$$

The theorem which is used in obtaining the expression for $\ln \Xi$ in terms of connected diagrams is the following:

THEOREM 1. Let N be a set of topologically different connected diagrams and M be the sum of all the connected and disconnected diagrams which are composed of arbitrary numbers of connected diagrams belonging to the set N. Then

{Sum of all the diagrams belonging to M}
= exp {Sum of all the diagrams belonging to N} -1.

Illustration of theorem 1. Take two connected diagrams $N \equiv z$ and fz^2. Construct M.

	z	z f z
N:	o,	o——o

M: 1) disconnected diagrams that contain one piece: o and o——o.

2) disconnected diagrams that contain two pieces: o—o and o o——o and o——o o——o. (Sym.

numbers 2, 1 and 2).

3) disconnected diagrams that contain three pieces: o o o (3!), o o o—o (2! 1!), o o—o o—o (1! 2!) and o—o o—o o—o (3!), etc.

PROOF. Let us assume that the set N consists of N diagrams $\gamma_1, \gamma_2, \ldots, \gamma_N$, where N may be an infinity. According to the hypothesis, any diagram belonging to M is uniquely characterized by the numbers of $\gamma_1, \gamma_2, \ldots \gamma_N$ from which it is built up, so that it will be written as $\Gamma(n_1, n_2, \ldots, n_N)$ where $n_1, n_2, \ldots$, and n_N denote the numbers of $\gamma_1, \gamma_2, \ldots$ and γ_N, respectively. The symmetry number of $\Gamma(n_1, n_2, \ldots, n_N)$ is

$$n_1!\ n_2! \ldots n_N! \prod_{i=1}^{N} [\text{symmetry number of } \gamma_i]^{n_i}.$$

Therefore we obtain

$$\Gamma(n_1, n_2, \ldots, n_N) = \frac{1}{n_1!\ n_2! \ldots n_N!} \gamma_1^{n_1} \gamma_2^{n_2} \ldots \gamma_N^{n_N}.$$

Summing up over all the sets of the integers including zero, over $n_1, n_2, \ldots$ and n_N, except for $n_1 = n_2 = \ldots = n_N = 0$, we obtain

$$\sum_{\{n_1, \ldots, n_N\}} \Gamma(n_1, n_2, \ldots, n_N) = \exp[\gamma_1 + \gamma_2 + \ldots + \gamma_N] - 1,$$

which is the relation to be proved.

We will give an explicite illustration of this theorem: Suppose again that we have only two types of connected diagrams: a black dot (representing the integral over z(r)) and two black dots connected with a line (representing the integral over $z(r_1)\ f(r_1, r_2)\ z(r_2)$). All disconnected diagrams which can be formed out of these two connected diagrams will consist of n_1 black dots and n_2 dumbells. The total sum over all connected and disconnected diagram is

$$\sum_{n_1=0}^{\infty} \sum_{n_1=0}^{\infty} [\underbrace{\bullet \bullet \bullet \ldots \bullet}_{n_1} + \underbrace{\bullet\!-\!\bullet\ \bullet\!-\!\bullet \ldots \bullet\!-\!\bullet}_{n_2}] - 1$$

or according to definition 2:

$$= \sum_{n_1=0}^{\infty} \sum_{n_2=0}^{\infty} \frac{1}{n_1!} \frac{1}{n_2!} \left[\int dr\ z(r) \right]^{n_1} \left[\frac{1}{2} \int dr_1 \int dr_2\ f(r_1 r_2) \right.$$

$$\left. z(r_1)\ z(r_2) \right]^{n_2} -1 = e^{\gamma_1} e^{\gamma_2} -1 = e^{[\bullet + \bullet\!-\!\bullet]} - 1$$

where

$$\gamma_1 = \int dr\ z(r)$$

$$\gamma_2 = \frac{1}{2} \int dr_1 \int dr_2\ f(r_1\ r_2)\ z(r_1)\ z(r_2)$$

Another theorem which is useful in taking partial summations or renormalization of diagrams is the following:

THEOREM 2. Let Γ be a diagram and N be a set of all the topologically different diagrams which are built up from Γ by replacing each part or element of Γ by one diagram, belonging to a set M. If each diagram can be reduced to Γ by replacing the parts of diagram by the original ones in Γ, uniquely, then

{Sum of all the diagrams belonging to N}
= A diagram which is obtained from Γ by taking the factor (corresponding to the parts which are replaced by those in M in getting N) to be the sum of the contributions from the diagrams in M.

(Partial summation or renormalization of diagrams.)

Example: Let the set of diagrams M consist of:

M:  ; 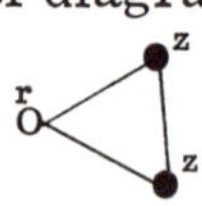 | $M \equiv \rho(r)$

that is

$$M(r) = \int f(r, r')\ z(r')\ dr' + \frac{1}{2} \int f(r\,r')\, f(r'\,r'')$$

$$f(r'\,r'')\ z(r')\ z(r'')\, dr'\ dr''.$$

And suppose the diagrams Γ is

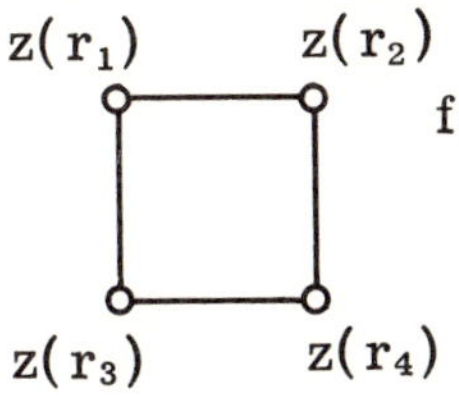

$$f(r_1 r_2) f(r_2 r_3) f(r_3 r_4) f(r_4 r_1) \times z(r_1) z(r_2) z(r_3) z(r_4)$$

Then the set of diagrams N are

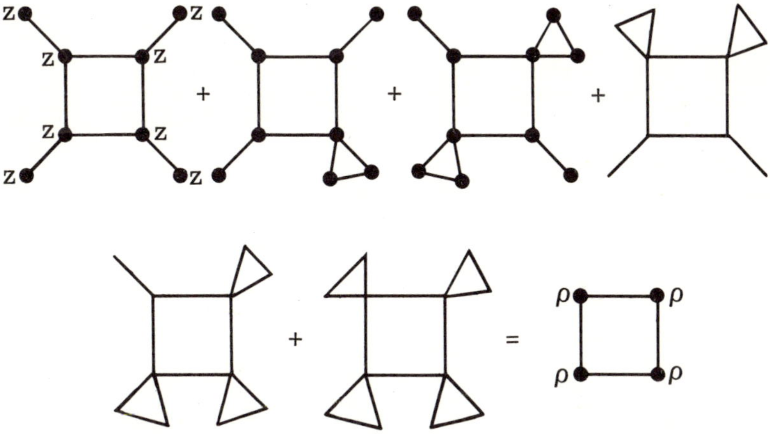

As a typical example, we shall consider the cluster expansion in the classical statistical mechanics. In the classical statistical mechanics, the grand partition function is given by

$$\Xi = \sum_{N=0}^{\infty} z^N Z_N, \tag{3.1}$$

$$Z_N = \frac{1}{N!} \int dr_1 \ldots \int dr_N \, e^{-\beta \sum_{i>j} v(|r_i - r_j|)} .$$

If we introduce b_{ij} by

$$b_{ij} = b(|r_i - r_j|) = \exp[-\beta v(|r_i - r_j|)] - 1, \tag{3.2}$$

then

$$\Xi = \sum_{N=0}^{\infty} \frac{1}{N!} z^N \int dr_1 \ldots \int dr_N \prod_{i>j} (1 + b_{ij})$$

$$= \sum_{N=0}^{\infty} \frac{1}{N!} z^N \int dr_1 \ldots \int dr_N \left\{ 1 + \sum_{i>j} b_{ij} + \Sigma\,\Sigma\, b_{ij} b_{kl} + \ldots \right\} \qquad (3.3)$$

By the aid of the Def. 2, this can be written as

Ξ = 1 + Sum of all the connected and disconnected diagrams which are products of arbitrary numbers of connected diagrams composed of $\overset{z}{\bullet}$ and $\underline{b}$ which connect two $\overset{z}{\bullet}$' s . (3. 4)

In the Mayer-series a term like $b_{ij}\, b_{ij}$ does not occur, hence there are no double bonds. If we compare this expression with theorem 1, we see at once that

ln Ξ = Sum of all the connected diagrams composed of $\overset{z}{\bullet}$'s. and $\overset{b}{-}$'s, each of which connects two $\overset{z}{\bullet}$'s. (3. 5)

Note that this quantity is extensive since every connected diagram gives a factor V. The total number of particles in this system is given by

$$N = \rho\Omega = \partial \ln \Xi / \partial \ln z,$$

ρ = z [1 + {Sum of all the connected diagrams which are composed of one white circle o , black circles $\overset{z}{\bullet}$ and bonds $\overset{b}{-}$}]. (3. 6)

Example of the functional derivative of a set of diagrams

lg Ξ =  + etc.

symmetry numbers 2 2 6.

$$\frac{N}{\Omega} = \partial \lg \Xi / \partial \lg z =$$

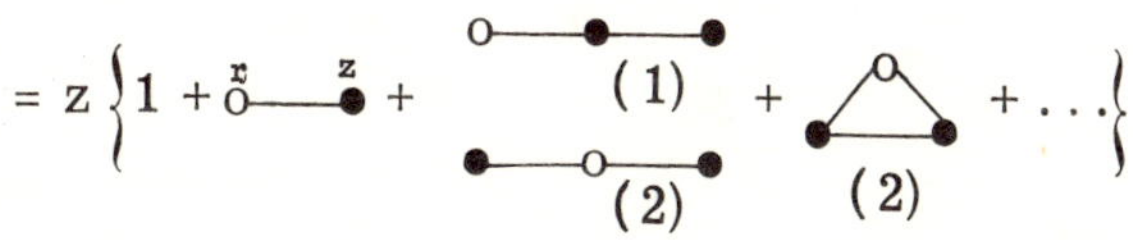

The third diagrams has a numerical factor $\frac{1}{2}$. After differentiation of z^3 the factor is 3/2, which is equal to 1 + 1/2, the sum of inverse symmetry numbers of the two resulting diagrams.

These diagrams are equivalent with those which are obtained from diagrams which have no articulation black circles in which the black circles are replaced by the diagrams for ρ given by (3. 6). Hence if we apply theorem 2, we get

$\rho = z$ [1 + {Sum of all the connected diagrams which are composed of one white circle o , black circles ρ● and bonds $\overset{b}{—}$, and which have no articulation black circles}]. (3. 7)

Example: Consider a diagram for ρ it will have one white circle and a number of black circles. Suppose one black circle is an articulation point, such as

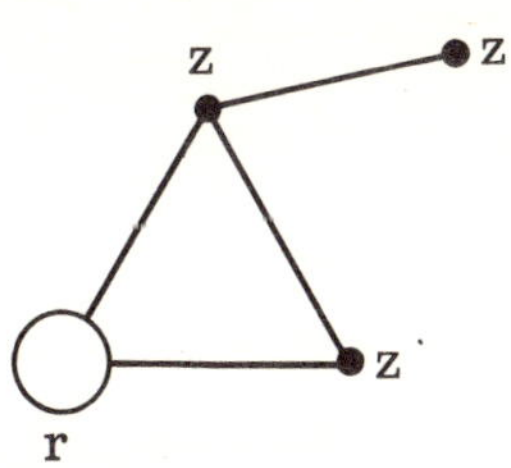

Let us combine this diagram with all those that have the same basic part (the triangle with the white dot) to which are attached all other diagrams of ρ:

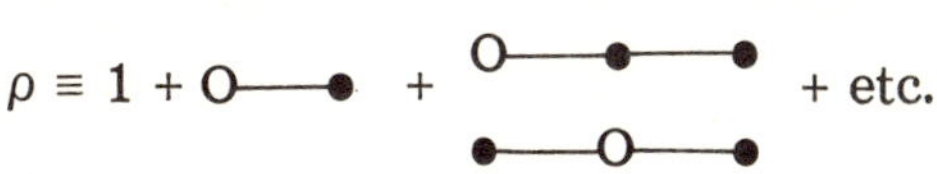

This is the sum:

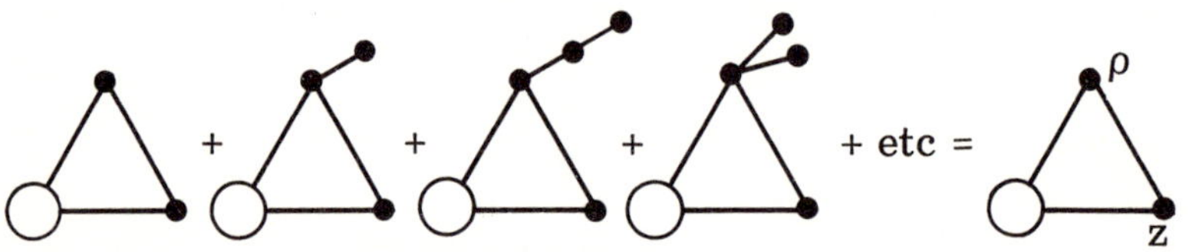

If we proceed in the same manner with all other points we will end up with connected diagrams without articulation points and labeled with ρ at the black circles.

If we consider that this sum of connected diagrams can also be taken as the sum of the connected and disconnected diagrams which are composed of arbitrary numbers of connected diagrams, each of which consists of one white circle o and black circles $\rho\bullet$ and bonds $\overset{b}{-}$ and which has no articulation circles. So far we have neglected the possibility that the white circle was an articulation circle. If this is so the diagram "comes apart". This is obvious if one writes out the integral.

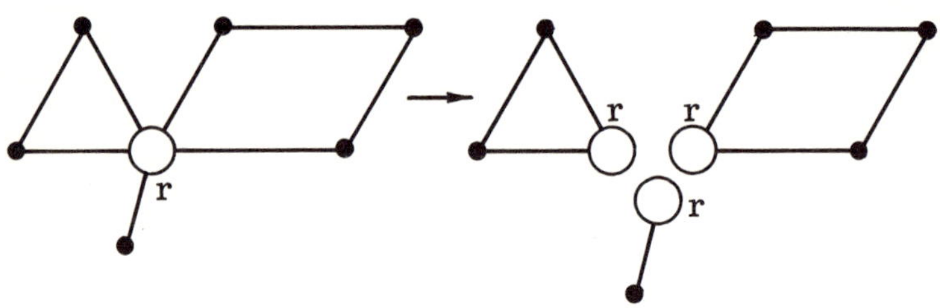

Hence we obtain disconnected diagrams without white articulation circles. (The steps in which the black articulation circles disappeared and the step in which the white articulation circles disappeared could have been taken in the opposite order just as well.) By an application of theorem 1, we get

$\rho = z \exp \{$ Sum of the connected diagrams which composed of one white circle o , black circles $\rho\bullet$ and bonds $\overset{b}{—}$ and which have no articulation circles$\}$, (3. 8)

or solving for ln z, we get

ln z = ln ρ - Sum of all the connected diagrams which are composed of one white circle o , black circles $\rho\bullet$ and bonds $\overset{b}{—}$ and which have no articulation circles. (3. 9)

In order to get the virial expansion formula for ln Ξ, that is the expansion in powers of density ρ, we have only to integrate the equation

$$\begin{aligned} \delta \ln \Xi &= \Omega\, \rho\, \delta \ln z \\ &= \delta[\Omega \rho \ln z] - \Omega \ln z\, \delta\rho \qquad (3.\,10) \\ &= \delta[\Omega\rho \ln z - \Omega\, [\ln \rho\, \delta\rho - \{\text{sum}\}\delta\rho] \end{aligned}$$

Upon substitution (3. 9) in the second term. Upon integration we find

$$\begin{aligned} \ln \Xi &= \Omega \rho \ln z - \Omega\, [\textstyle\int \ln \rho\, d\rho - \int \{\text{sum}\}\, d\rho] \\ &= \Omega\, \rho \ln z - \Omega\, [\rho \ln \rho - \rho - \textstyle\int \{\text{sum}\}\, d\rho] \end{aligned}$$

$\ln \Xi = \Omega\rho[\ln z - \ln \rho + 1]$

$+ \{$Sum of all the more than singly connected diagrams which are composed of black circles $\rho\bullet$ and bonds $\overset{b}{—}$.$\}$. (3. 11)

The integration constant has been determined by the condition that ln Ξ ρ are zero when z = 0.

4. QUANTUM STATISTICAL MECHANICS. 1.

The Hamiltonian of a system of N particles is taken to be

$$H_N = K_N + V_N, \tag{4. 1}$$

where K_N and V_N stand for the kinetic and potential energy parts which are usually taken to be

$$K_N = -\frac{1}{2m}\sum_{i=1}^{N}\frac{\partial^2}{\partial r_i^2}, \quad V_N = \sum_{N\geq i>j\geq 1} v(|r_i - r_j|) \tag{4. 1'}$$

in the representation where particles are taken to be distinguishable.

In order to investigate the thermodynamic properties, we shall investigate the logarithm of the grand partition function which is the most convenient free energy for the diagrammatical analysis. The grand partition function as a function of its proper independent variables, volume Ω, fugacity z and inverse temperature β, is given by

$$\Xi = \sum_{N=0}^{\infty} z^N Z_N, \quad Z_N = \mathrm{tr}_{(N)}\, e^{-\beta H_N} \tag{4. 2}$$

As already mentioned in this series of lectures by Meijer and Tanaka, the grand partition function can be calculated either in the representation where particles are distinguishable by the aid of the Hamiltonian (4. 1) or in the representation of the second quantization where particles are indistinguishable. Of course, they give the same things. But it is easier for us to visualize the assembly of distinguishable particles than the assembly of indistinguishable particles. And hence we shall do the former first in details and then the latter later, in the next section.

In the momentum representation, the complete set of one particle wave functions is taken to be

$$\varphi_k(r) = \frac{1}{\sqrt{\Omega}}\, e^{ik\cdot r} \tag{4. 3}$$

The complete set of N distinguishable particle wave functions is taken to be.

$$|k_1, k_2, \ldots, k_N\rangle \equiv \varphi_{k_1}(r_1)\ \varphi_{k_2}(r_2)\ldots\ \varphi_{k_N}(\gamma_N). \quad (4.4)$$

The symmetrized complete set of N indistinguishable particles is

$$|\psi_{k_1}, \ldots, {}_{k_N}\rangle = \frac{1}{\sqrt{N!}} \sum_P^{(N!)} \epsilon^P |k_1^P, \ldots, k_N^P\rangle, \quad k_1 \leq \ldots \leq k_N. \quad (4.5)$$

The normalization is

$$\langle \psi_{k_1}, \ldots, {}_{k_N} | \psi_{k_1}, \ldots, {}_{k_N}\rangle = \Pi_k\ n_k\,!, \quad (4.6)$$

where n_k is the number of k among $k_1, k_2, \ldots, k_N$: $n_k = \Sigma_{i=1}^{N} \delta_{k, k_i}$, and $k_1^P, \ldots, k_N^P$ is some permutation of $k_1, \ldots, k_N$. For instance, k_1^P, k_2^P is either k_1, k_2 or k_2, k_1. Next we shall define $\psi_{k}, \ldots, {}_{k_N}$ for an arbitrary order of $k_1, \ldots, k_N$ by

$$|\psi_{k_1}, \ldots, {}_{k_N}\rangle = \epsilon^P |\psi_{k_1^P}, \ldots, {}_{k_N^P}\rangle. \quad (4.7)$$

Then the trace of a quantity A_N is given by

$$\mathrm{tr}_{(N)} A_N = \sum_{k_1 \leq \cdots \leq k_n} \cdots \sum \frac{1}{\Pi_k n_k!} \langle \psi_{k_1}, \ldots, {}_{k_N} | A_N | \psi_{k_1}, \ldots, {}_{k_N}\rangle$$

(indistinguishable particles)

$$= \frac{1}{N!} \sum_{k_1} \cdots \sum_{k_N} \langle \psi_{k_1}, \ldots, {}_{k_N} | A_N | \psi_{k_1}, \ldots, {}_{k_N}\rangle.$$

(distinguishable particles)

If A_N is a quantity which is symmetric about the coordinates of N particles that is $A_N P = PA_N$, then this reduces to

$$\mathrm{tr}_{(N)} A_N = \frac{1}{N!} \sum_{k_1} \cdots \sum_{k_N} \sum^{(N!)} \epsilon^P \langle k_1, \ldots, k_N^P | A_N | k_1, \ldots, k_N\rangle. \quad (4.8)$$

In order to calculate the grand partition function, we have to calculate the trace of $\exp -\beta H_N$ which is expanded in powers of V_N as

$$e^{-\beta H_N} = \sum_{n=0}^{\infty} (-1)^n \int_0^{\beta} du_1 \int_0^{u_1} du_2 \ldots \int_0^{u_{n-1}} du_n \, e^{-(\beta-u_1)K_N} V_N$$

$$e^{-(u_1-u_2)K_N} V_N \ \ldots \ V_N \, e^{-(u_{n-1}-u_n)K_N} V_N \, e^{-u_n K_N}. \qquad (4.9)$$

Let us introduce unity

$$1 = \sum_{k_1} \cdots \sum_{k_N} |k_1, \ldots, k_N \rangle \langle k_1, \ldots, k_N| \qquad (4.10)$$

which is the consequence of the completeness of (4.4), between every successive pairs of operators in (4.9). Then we have

$$\Xi = \sum_{N=0}^{\infty} \frac{1}{N!} z^N \sum_{k_1} \cdots \sum_{k_N} \sum_{P}^{(N!)} \epsilon^P \langle k_1^P, \ldots, k_N^P | e^{-\beta H_N} | k_1, \ldots k_N \rangle \qquad (4.11)$$

$$\langle k_1^P, \ldots, k_N^P | e^{-\beta H_n} | k_1, \ldots, k_N \rangle =$$

$$= \sum_{i_1 > j_1} \cdots \sum_{i_n > j_n} \int_0^{\beta} du_1 \cdots \int_0^{u_{n-1}} du_n$$

$$\Sigma_{q_1} \ldots \Sigma_{q_N} e^{-(\beta-u_1) \sum_{i=1}^{N} \epsilon_{k_i}} V_{q_1} e^{-(u_1-u_2) \Sigma \epsilon_{k_i}^{(2)}} V_{q_2} \ldots,$$

$$\ldots V_{q_{n-1}} e^{-(u_{n-1}-u_n) \Sigma \epsilon_{k_i}^{(n-1)}} V_{q_n} e^{-u_n \Sigma \epsilon_{k_i}} \qquad (4.11')$$

$$\epsilon_{k_i}^{(n)} = \epsilon_{k_i}, \quad \epsilon_{k_i}^{(v-1)} = \epsilon_{k_i}^{(v)} + q_1(\delta_{i,iv} - \delta_{j,jv})$$

We can write the integrand by the aid of the diagram as follows. [Cf. Fig. 4-1.] We shall draw N (vertical) solid lines which start at $u = 0$ with the states $k_1, \ldots, k_N$ and end at $u = \beta$ with the states $k_1^P, \ldots, k_N^P$, respectively, and draw n (horizontal) dotted lines connecting two solid lines, i_1 th and j_1 th, . . . , i_n th and j_n th, respectively, at time $u_1, \ldots, u_n$ and then connect 1st, . . . , Nth lines at $u = 0$ and P(1st, 2nd, . . . , Nth) lines at $u = \beta$. In this

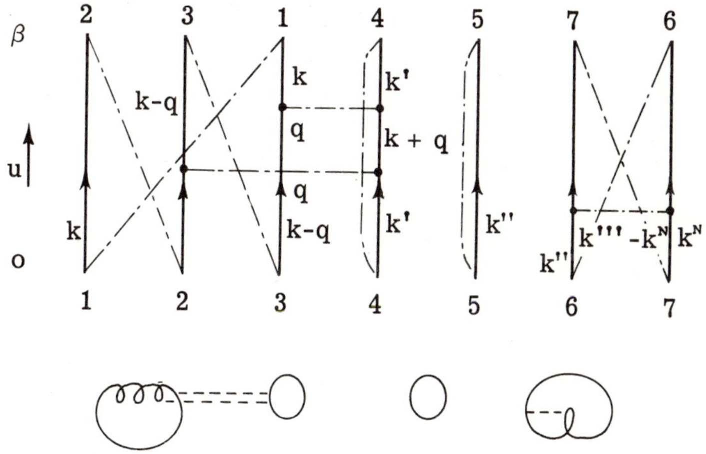

Fig. 4-1. An example of diagram for the integrand of (4. 11').

way we get a diagram which is composed of closed lines which may pass the $0 - \beta$ line several lines and dotted lines which connect two solid lines. The closed solid lines are called "torons" and especially the one which passes the $0 - \beta$ line n times an "n-toron " by Montroll and Ward. It is noticed here that N is the total number of particles and is equal to the number of solid lines which pass the $0 - \beta$ line and hence we shall associate a factor z to the part of solid line at the $0 - \beta$ line. The number of transpositions for an "n-toron" is shown to be equal to n - 1 and ϵ^P can be taken a product of these factors, ϵ^{n-1}, each from a toron. Taking these facts into account, we find that

Ξ = 1 + Sum of all the connected and disconnected diagrams which are composed of arbitrary numbers of connected diagrams composed of torons and dotted lines connecting torons. (4. 12)

The factors to be associated to elements of the diagram are

1) $\exp -(u - v)\Sigma_k$ for a solid line of state k, starting at v and ending at u.
2) $-v_q$ for a dotted line connecting two solid lines, changing them from k and k' to k + q and k' - q.
3) $\epsilon^{n-1} a^n$ for an "n-toron."

The integrations and summations to be performed are over all times u and momentum transfers q of interactions and the initial states, $k_1, \ldots, k_N$.

Applying the theorem 1, we get at once.

$\ln \Xi$ = Sum of all the connected diagrams which are composed of torons and dotted lines connecting torons. (4. 13)

This is the expression of the logarithm of the grand partition function by means of the diagrams by Montroll and Ward.

The diagrams which have no dotted (interaction) lines among (4. 13) are those for the ideal gas and are summed up to

$$\Xi_0 = \Pi_k[1 - \epsilon z e^{-\beta\epsilon_k}]^{-\epsilon},$$

$$\ln \Xi_0 = -\epsilon \sum_k \ln [1 - \epsilon z e^{-\beta\epsilon_k}]. \tag{4. 14}$$

The diagrams which have interaction lines among (4. 13) can be considered to be obtained from the diagrams which are drawn between 0 and β allowing solid lines with increasing and decreasing time by replacing a solid line starting at a time v at an interaction and ending at a later/earlier time u at another interaction by solid lines starting at the time v and ending at the time u encircling the 0 - β line arbitrary times from zero time once to infinite times; cf. Fig. 4-2. Thence, applying the Theorem 2, we get

$$\ln \Xi = -\epsilon \sum_k \ln [1 - \epsilon z e^{-\beta\epsilon_k}]$$

+ Sum of all the connected diagrams which are composed of loops of solid lines with increasing and decreasing time and dotted (interaction) lines connecting them. (4.15)

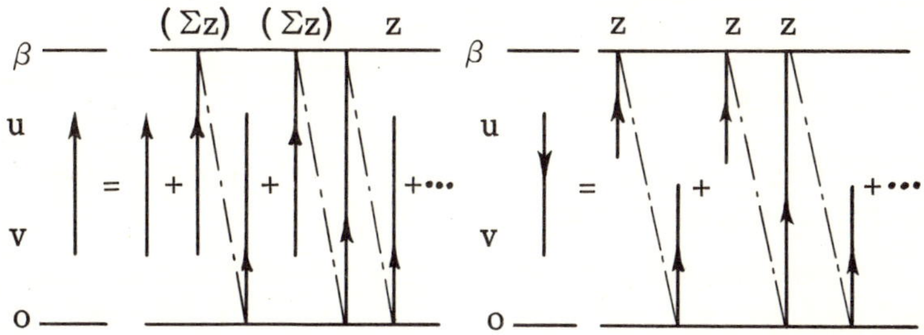

Fig. 4 - 2.

The factor which is to be associated to the solid line which starts at time v and ends at a later/earlier time u is

$$e^{-(u-v)\epsilon_k} / [1 - \epsilon z e^{-\beta\epsilon_k}],$$

$$\epsilon z e^{-(\beta\epsilon - v + u)\epsilon_k} / [1 - \epsilon z e^{-\beta\epsilon_k}],$$

respectively. An additional factor ϵ should be associated to each loop of solid lines. Several terms are illustrated in Fig. 4-3.

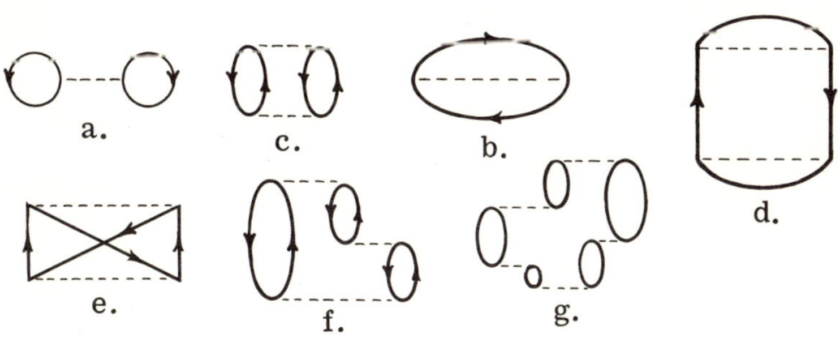

Fig. 4 - 3.

a. The direct term in the first order perturbation. The diagrams which have a dotted line connecting two separate parts are unique in the point that it depends on the value of $v_0 = \int v(r)\,dr$; cf., the discussion under (5. 1).

b. The exchange term.

c. f and g. The ring terms which are evaluated by Montroll and Ward for an electron gas.

d. An anomalous term which has been specially analyzed by Luttinger and Ward, in their consideration of the limit of absolute zero. This diagram is special in that the upward and downward lines should be of the same momentum.

Discussion of figure 4-3.

a 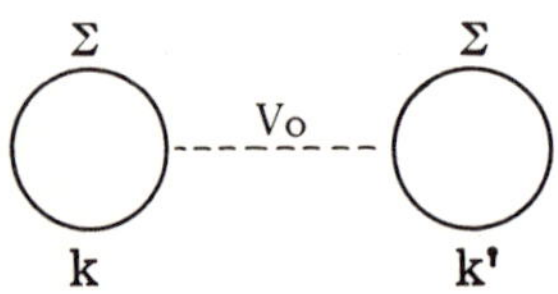

$q = 0$ no transfer of momentum

The factor associated with this diagram is

$$\frac{1}{2}\sum_{k}\sum_{k'} \epsilon^2 \frac{\epsilon z e^{-\beta\epsilon_k}}{1-\epsilon z e^{-\beta\epsilon_k}} \frac{\epsilon z e^{-\beta\epsilon_{k'}}}{1-\epsilon z e^{-\beta\epsilon_{k'}}} \frac{v_0}{\Omega}$$

↑ symmetry number.

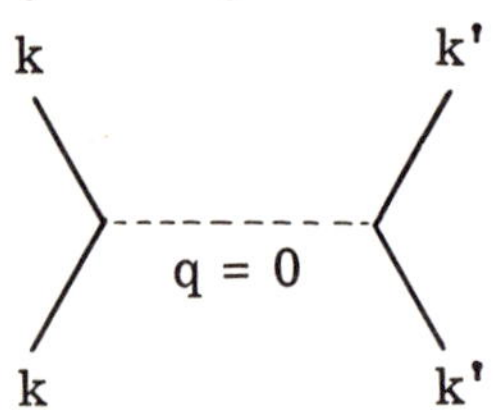

Both lines are going downwards, hence ϵz in the numerator. They are actually:

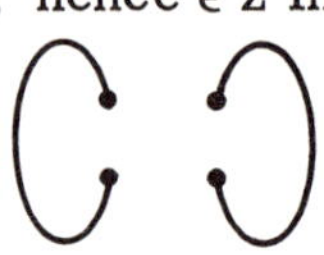

b

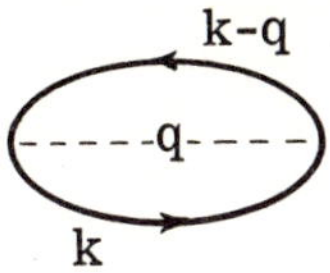

$$\frac{1}{2} \sum_k \sum_q \frac{\epsilon z e^{-\beta\epsilon_k}}{1 - \epsilon z e^{-\beta\epsilon_k}} \frac{\epsilon z e^{-\beta\epsilon_{k-q}}}{1 - \epsilon z e^{-\beta\epsilon_{k-q}}} \frac{v_q}{\Omega}$$

c

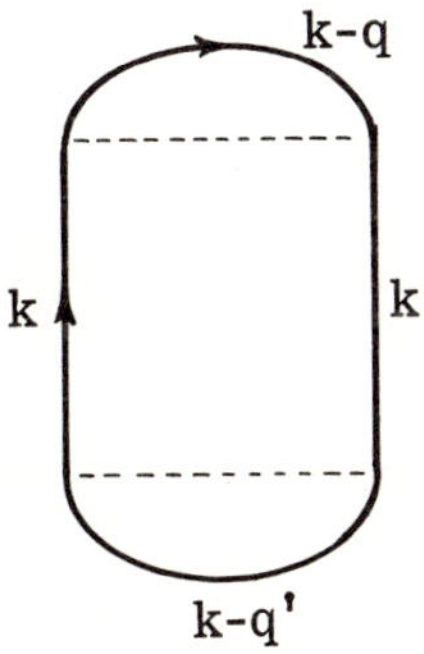

All other double sums contain terms in which k = k', but here we have this term only.

$$\overset{\text{up line}}{\frac{e^{-(u-v)\epsilon_k}}{1 - \epsilon z e^{-\beta\epsilon_k}}} \quad \overset{\text{down line}}{\frac{e^{(u-v)\epsilon_k} \epsilon z e^{\epsilon_k}}{1 - \epsilon z e^{-\beta\epsilon_k}}} =$$

Hence product:

$$\frac{1}{1 - \epsilon z e^-} \times \frac{\epsilon z e^-}{1 - \epsilon z e^-} = -(1 - f) f.$$

Fermion $\epsilon = -1$

$$f = \frac{-z e^-}{1 + z e^-} = \frac{-1}{z e^- + 1}$$

This goes to zero for $\beta \to 0$.*

* The solution of the riddle is that if temperature is introduced as a perturbation, this set of terms has no "starting" point.

However in the expression

$$\frac{1}{1 - z\, e^{-\beta\epsilon_k}}\,, \qquad z = e^{\beta\mu} \quad \text{where } \mu \text{ is}$$

The Fermi energy in the unperturbed system. Analysis by Luttinger and Ward shows however that with interaction μ is different. Hence the diagrams do not dissappear for $T \to 0$ (even with infinitesimal interaction).

5. QUANTUM STATISTICAL MECHANICS. 2

The result obtained, in the last section, (4. 15), is to be compared with the results which have been indicated at the end of Meijer's lecture. Actually this result can be obtained more properly in the formalism of the second-quantization. This will be undertaken in the first part of this section.

The Hamiltonian in the formalism of the second-quantization, is given by

$$H = H_0 + V, \tag{5. 1}$$

$$H_0 = \Sigma_k \epsilon_k a_k^+ a_k\,,$$

$$V = \frac{1}{\Omega} \Sigma_k \Sigma_{k'} \Sigma'_q v_q a_{k+q}^+ a_{k'-q}^+ a_{k'} a_k\,,$$

where

$$\epsilon_k = k^2/2m \qquad \text{and} \qquad v_q = \int dr\, e^{ig.r}\, v(r)\,,$$

and the prime on Σ'_q means that the summation is taken over half of q, e. g., over q for which $q_x > 0$. We shall take $v_0 = 0$, without loss of generality in the following. If v_0 is not zero, it appears in the Hamiltonian as an additive constant $\frac{1}{2}\frac{N^2}{\Omega} v_0$, which is to be added to the final expression for the energy and hence to the free energy. This correction results in an additive term $1/2\ \beta N^2/\Omega\ v_0$ to the logarithm of the grand partition function. * **

---

* We shall give some relations between corresponding quantities for systems of which Hamiltonians are H_N and H_N^*, which are related by

$$H_N = H_N^* + f(N) \tag{5.2}$$

where $f(N)$ is a function only of N and Ω. The quantities for two systems are discriminated by a star. The partition functions are related by

$$Z_N = \mathrm{tr}_{(N)} e^{-\beta H_N} = \mathrm{tr}_{(N)} e^{-\beta H_N^*} e^{-\beta f(N)} = Z_N^* \, e^{-\beta f(N)}$$

$$\therefore \ \ln Z_N = \ln Z_N^* - \beta f(N) \tag{5.3}$$

The chemical potentials are related by

$$\ln z = -\frac{\partial}{\partial N} \ln Z_N = -\frac{\partial}{\partial N} \ln Z_N^* + \beta \frac{\partial}{\partial N} f(N)$$

$$\ln z = \ln z^* + \beta \frac{\partial}{\partial N} f(N) \tag{5.3'}$$

Hence

$$\ln \Xi = N \ln z + \ln Z_N = N \ln z^* + N \beta \frac{\partial}{\partial N} f(N) + \ln Z_N^* - \beta f(N)$$

$$\ln \Xi = \ln \Xi^* + \beta \left\{ N \frac{\partial}{\partial N} f(N) - f(N) \right\} \tag{5.3''}$$

For $f(N) = \frac{1}{2} N^2 v_0$, this additive term is $\beta \frac{1}{2} N^2 v_0$.

** In this passage, it is noticed that, for a neutral system composed of charged particles, we can start by assuming $v_{0,ij}$, the zero-component of Fourier transform of potential energy between two charged particles, of species i and j, to be zero, without loss of generality. This fact is proved as follows. It is noticed that $v_{0,ij}$ appear only in the additive term of the Hamiltonian in the form:

$$V_0 = \frac{1}{2\Omega} \sum_i{}' N_i^2 \, v_{0,ii} + \frac{1}{\Omega} \sum_{i>j} N_i N_j \, v_{0,ij} \, . \tag{5.4}$$

where N_i is the number of particles in the system, and $v_{0,ij}$ is proportional to the charges of particles of ith and jth species and hence we can write $v_{0\,ij} = q_i q_j v_0$. Substituting this into (5.4), we get

$$V_0 = \frac{1}{2} (\Sigma_i N_i q_i)^2 \; v_0 \tag{5 4'}$$

which is zero for a neutral system. This can be obtained by; taking $v_{0,ij} = 0$ and there exist no other places where $v_{0,ij}$ appear and hence we can take $v_{0,ij}$ to be zero without loss of generality for a neutral system composed of charged particles.

The operators a_k and a_k^+ satisfy the commutation relations:

$$a_k a_{k'}^+ - \epsilon\, a_{k'}^+ a_k = \delta_{k,k'}, \tag{5.5}$$

$$a_k a_{k'} - \epsilon\, a_{k'} a_k = 0.$$

This Hamiltonian is obtained from the usual one in the coordinate representation, by introducing the transformation:

$$\psi(r) = \frac{1}{\sqrt{\Omega}} \Sigma_k e^{ik.r} a_k \tag{5.6}$$

The above form is proper for our development because H_0 is diagonal in this representation. From the commutation relations (5.5), we have

$$a_k^+(u) = e^{u\epsilon_k} a_k^+ \qquad \text{and} \qquad a_k(u) = e^{-u\epsilon_k} a_k \tag{5.5'}$$

and hence

$$v(u) = e^{uH_0} v\, e^{-uH_0}$$

$$= \Sigma'_q \Sigma_k \Sigma_{k'} e^{u(\epsilon_{k'+q}+\epsilon_{k-q}-\epsilon_k-\epsilon_{k'})} a_{k'+q}^+ a_{k-q}^+ a_k a_{k'} \tag{5.7}$$

The grand partition function is given by

$$\Xi = \mathrm{tr}\, z^N e^{-\beta H} \tag{5.8}$$

where $N = \Sigma_k a_k^+ a_k$. (5.9)

By expanding exp $-\beta$ H in the integrand in powers of V, we get

$$\Xi/\Xi_0 = \int_0^{\beta} du_1 \int_0^{u_1} du_2 \cdots \int_0^{u_{n-1}} du_n \langle v(u_1)\, v(u_2) \cdots v(u_n) \rangle_0$$

where

$$\langle A \rangle_0 = \mathrm{tr}(A\, z^N e^{-\beta H_0})\, /\Xi_0, \tag{5.11}$$

$$\Xi_0 = \mathrm{tr}\, z^N e^{-\beta H_0} = (4.14)$$

By the aid of (5. 7), we get

$$\Xi/\Xi_0 = \int_0^\beta du_1 \, \Sigma'_{q_1} \cdots \Sigma'_{q_n} v_{q_1} \cdots v_{q_n} \Sigma_{k_1} \Sigma'_{k_1} \cdots \Sigma_{k_n} \Sigma'_{k_n}$$

$$\times e^{u_1 (\epsilon_{k'_1+q_1} + \epsilon_{k_1-q_1} - \epsilon_{k_1} - \epsilon_{k'_1})} \cdots e^{u_n (\epsilon_{k'_n+q_n} + \epsilon_{k_n-q_n} - \epsilon_{k_n} - \epsilon_{k'_n})}$$

$$\times \langle a^+_{k'_1+q_1} a^+_{k_1-q_1} a_{k_1} a_{k'_1} \cdots a^+_{k'_n+q_n} a_{k_n} a_{k'_n} \rangle_0 . \quad (5.12)$$

In order to get the diagrammatical expression, we shall apply Wick's theorem for the quantum statistics:

$$\langle AB \cdots YZ \rangle_0 = \Sigma^+_P \, \epsilon \, \langle A^P B^P \rangle \langle C^P D^P \rangle \cdots \langle Y^P Z^P \rangle \quad (5.13)$$

where

$$\langle a^+_k a_{k'} \rangle_0 = \delta_{kk'} \frac{z e^{-\beta \epsilon_k}}{1 - \epsilon z e^{-\beta \epsilon_k}},$$

$$\langle a^+_k a_{k'} \rangle_0 = \delta_{kk'} \frac{1}{1 - \epsilon z e^{-\beta \epsilon_k}}, \quad \langle a_k a_{k'} \rangle = 0 \quad (5.14)$$

and Σ^+_P means that the summation is to be taken over all the permutations by which the orders of the operators in the pairs A^P and B^P, C^P and D^P, ... and also the order of A^P, C^P, ... Y^P are the same as in the original product: ABC ... YZ. This theorem is proved, for instance, by the mathematical induction:

1) For a product of pair of operators, we have only the identity permutation and hence (5. 13) is true.
2) Now let us consider a product of 2n operators. The first element is either a creation operator a^+_k or an annihilation operator a_k, we shall treat both cases by writing them as $a^\pm_k$ (the lower sign correspond to an annihilation operator). So that we can write the average of a product of 2n operators as

$$\langle a^\pm_k BC \cdots YZ \rangle_0 . \quad (5.15)$$

If we use (5. 5'), we can see easily that

$$<BC \cdots YZ\, a_k^{\pm}>_0 = z^{\mp}\, e^{\pm\beta\epsilon_k} <a_k^{\pm} BC \cdots YZ>_0$$

Hence

$$<a_k^{\pm}\, BC \cdots YZ>_0 = \frac{1}{1 - \epsilon z^{\mp} e^{\pm\beta\epsilon_k}} \times$$

$$\times\, [<a_k^{\pm}\, BC \cdots z>_0 - \epsilon <BC \cdots Z a_k^{\pm}>_0]$$

$$= \frac{1}{1 - \epsilon z^{\mp} e^{\pm\beta\epsilon_k}} \{<[a_k^{\pm}\, B]_\epsilon C \cdots Z>_0 + \epsilon <B[a_k^{+}\, c]_\epsilon \cdots Z>_0$$

$$+ \cdots + \epsilon <BC \cdots [a_k^{+}\, Z]_\epsilon>_0 \quad (5.17)$$

where $[A, B]_\epsilon = AB - \epsilon BA$. If we notice that

$$\frac{[a_k^{\pm}\, B]}{1 - \epsilon Z^{\mp} e^{\pm\beta\epsilon_k}} = <a_k^{\pm}\, B>_0. \qquad (5.18)$$

We find that

$$<a_k^{\pm}\, BC \cdots Z>_0 = <a_k^{\pm}\, B>_0 <C \cdots Z>_0 + \epsilon <a_k^{\pm}\, C>_0 <BC \cdots Z>_0$$

$$+ \cdots + \epsilon <a_k^{\pm} Z>_0 <BC \cdots Y>_0. \qquad (5.19)$$

It is noticed that $a_k^{\pm}$ is paired with all of B, ... and Z and ϵ appears for the odd permutation carrying the operator paired with $a_k^{\pm}$ to the next of $a_k^{\pm}$. This identity shows that if the theorem is true for the average of any product of 2(n - 1) operators, it is also true for any product of 2n operators. Thus the theorem is proved.

We shall take the direction of time vertically from 0 to β. Corresponding to an interaction at u_1, we shall write a dotted horizontal line which connect two solid lines incoming with states k_1/k_1' corresponding to two annihilation operators and outgoing with state $k_1 - q_1/k_1' - q_1$ corresponding to the creation operators at time u_1. We shall draw an arrow in such a way that the conservation of momentum is clarified, on the dotted line. Similarly we shall write dotted lines connecting solid lines at $u_2, \ldots$ and u_n. Corresponding to each way of the pairing, we shall draw a diagram. We shall connect solid lines which are made to correspond to creation and annihilation operators which are paired. The factors which are due to the interaction at u_1 are v_{q_1}/Ω and $\exp u_1 (\epsilon_{k'_1+q_1} + \epsilon_{k_1-q_1} - \epsilon_{k'_1} - \epsilon_{k_1})$ and the factor to be associated to the solid line due to pairing is either $\langle a\, a^+\rangle_0$ or $\langle a^+\, a\rangle_0$. We can associate v_{q_1}/Ω only to the dotted line, and exponentials and $\langle a\, a^+\rangle$ or $\langle a^+\, a\rangle$ to the solid lines according to their momentum. Then the factors to be associated are exponentials from the starting point v and ending point u and (5. 14); the product is equal to (5. 3), except for the sign factor ϵ for the case of $v > u$. Next let us consider the sign ϵ^P. As stated above, the solid lines make loops. We shall collect the creation and annihilation operators which belong to a loop, without changing the order of operators belonging to the same loop. The permutation has an even parity for that process. An example of the product of operators for a loop is given in Fig. 5-1. An a^+ and the next a represent a pair of

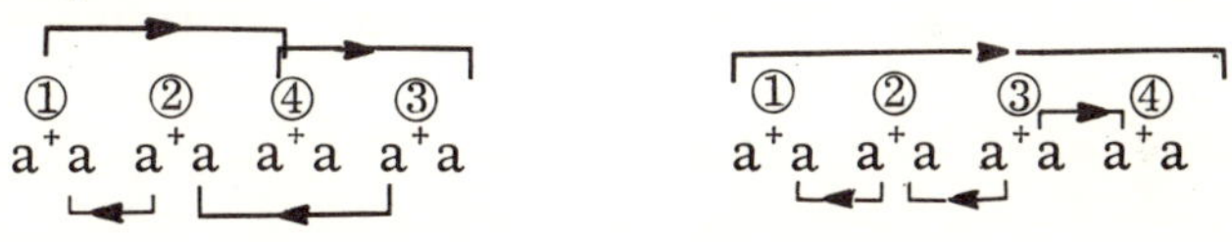

Fig. 5 - 1. An example of the product of operators for a loop.

operators associated to an end of a dotted line. It is shown that the parity for a loop is odd or even according as the number of downward solid lines is even or odd. This can be as follows. As an example, we shall consider Fig. 5 - 1, where we have shown the pairs to be paired or contracted, by connecting them by a solid line with an arrow, starting from a^+ and ending at a. We shall call the most left set as the 1st and the set of which a^+ is paired or contracted with a of nth set to be $(n + 1)$st. Let us rearrange the product in such a way that nth set to the most right, (n - 1) st to the next and ... and 1st one at the most left. Then the contracted pairs are ordered in such a way that a is the left of a^{+} except one. This is alright for $\underleftarrow{a\ a^{\pm}}$ is the original order, but not for $\overrightarrow{a^{+}\ a}$ was the original order except for the contraction of a^{+} of the 1st and a for the nth set; cf., the directions of arrows in both diagrams in Fig. 5 - 1. So that in order to carry them without changing their original order, we have to make a transpositions, of which times is equal to the number of pairings or contractions of the type $\overrightarrow{a^{+}\ a}$ minus one for a loop.

The diagrams obtained will generally consist of several separate connected parts. For instance, let us consider diagrams which consist of n_1 connected parts of type γ_1 made of ν_1 dotted lines, n_2 of the type γ_2 made of ν_2 dotted lines, ... The number of such diagrams is equal to

$$\frac{n!}{n_1!\, n_2! \cdots (\nu_1 !)^{n_1} (\nu_2 !)^{n_2} \cdots}.$$

These are different only in the relative times of interactions with each other. If the time dependence of quantities in a connected part does not change by the relative times composed with the other connected parts, we can integrate over time variables. If we consider

$$\frac{n!}{(\nu_1)^{n_1} (\nu_2)^{n_2}} \cdots$$

separate diagrams we can integrate. If we did that we have included the same diagrams $n_1!\, n_2! \ldots$ times. Cf. Fig. 5-2. Hence we get

$$\Xi/\Xi_0 = \sum \frac{1}{n_1!\, n_2! \cdots} \gamma_1^{n_1} \gamma_2^{n_2} \cdots = \exp(\gamma_1 + \gamma_2 + \ldots)$$

or

$$\ln \Xi = \ln \Xi_0 + \gamma_1 + \gamma_2 + \ldots$$

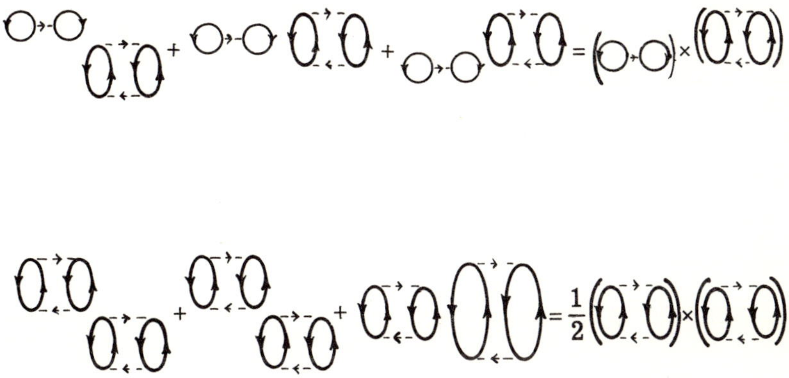

Fig. 5 - 2. Examples of integrals in reducing the disconnected parts into independent integrals.

$\ln \Xi = \ln \Xi_0$ + Sum of all the connected diagrams which are composed of

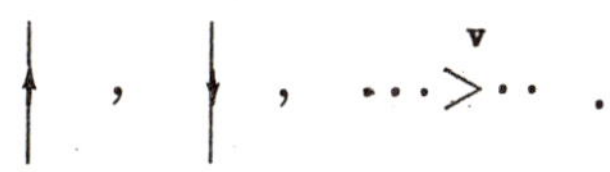 (5. 21)

If we define

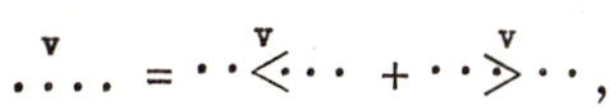 (5. 22)

and take account of theorem 2, we find

$\ln \Xi = \ln \Xi_0$ + Sum of all the connected diagrams composed of

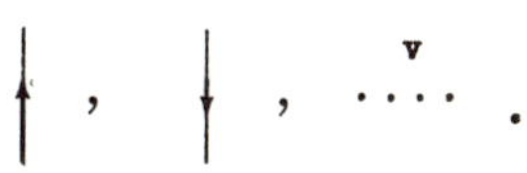 (5. 23)

which is identical with (4. 15).

In the above, we considered the direction of lines and distinguished the upward and downward lines. But we can take them indistinguishable and understand that the integral at any interaction be taken but the factor to be associated with any propagator is (5.14) according as the times for the starting point of that propagator is later or earlier than its ending point. We shall accept this premise and here we shall take attention to the identity:

$$-\sum_{l=-\infty}^{\infty} e^{-(u-v)\zeta_l} \frac{1}{\zeta_l - \epsilon_k} = \begin{cases} e^{-(u-v)\epsilon_k} \dfrac{1}{1-\epsilon z e^{-\beta\epsilon_k}}, & u > v \\ e^{-(u-v)\epsilon_k} \dfrac{\epsilon z e^{-\beta\epsilon_k}}{1-\epsilon z e^{-\beta\epsilon_k}}, & v > u \end{cases} \tag{5. 24}$$

where

$$\zeta_l = (2l+1)\, i\pi/\beta + \ln z/\beta \qquad \text{for fermion.}$$

$$\zeta_l = 2l\, i\pi/\beta + \ln z/\beta \qquad \text{for boson.}$$

This is confirmed as follows. Taking attention to the fact that

$$-\beta \frac{1}{1-\epsilon z e^{-\beta\zeta}} \quad \text{or} \quad -\beta \frac{\epsilon z e^{-\beta\zeta}}{1-\epsilon z e^{-\beta\zeta}}$$

has poles at ζ_l's and the residues at these points are unity, we shall rewrite the lefthand side by the aid of a contour integral:

$$\beta \times \frac{1}{2\pi i} \int_c d\zeta \frac{1}{1-\epsilon z e^{-\beta\zeta}} e^{-(u-v)\zeta} \frac{1}{\zeta - \epsilon_k} \tag{5. 25a}$$

or

$$\beta \times \frac{1}{2\pi i} \int_c d\zeta \frac{\epsilon z e^{-\beta\zeta}}{1-\epsilon z e^{-\beta\zeta}} e^{-(u-v)\zeta} \frac{1}{\zeta - \epsilon_k} \tag{5. 25b}$$

C is taken to include all the ζ_l's. We shall deform the contour C to the difference of the circle of infinite radius, C_2, and the small circle around ϵ_k, C_1; then

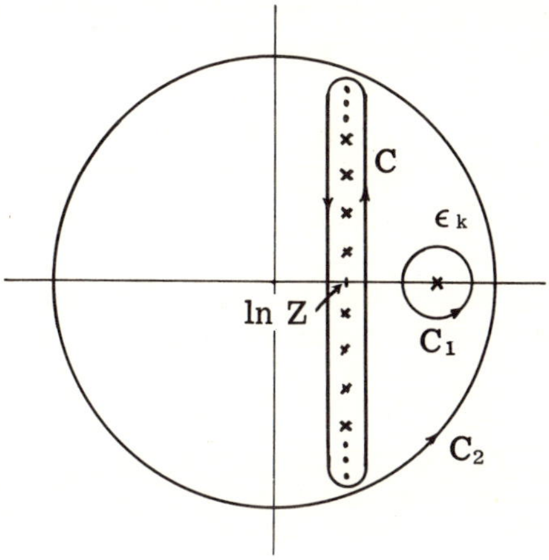

Fig. 5-3. Contours C, C_2 and C_1.

$$\int_c = \int_{c_2} - \int_{c_1} .$$

Cf. Fig. 5-2. The contour integral of (5.25a) and (5.25b) along C_2 are zero when $u > v$ and $\beta - v + u > 0$, respectively. And we get the righthand sides of (5.24) from the contour integral along C_1, for the respective cases.

Substituting this closed expression for $< a^{\pm} a^{\mp} >_0$, we can integrate over time for any interaction, finding

$$\beta\, \delta^{\zeta_{l_1}+\zeta_{l_2}-\zeta_{l_3}-\zeta_{l_4}} = \int_0^\beta du\, e^{-u(\zeta_{l_1}+\zeta_{l_2}-\zeta_{l_3}-\zeta_{l_4})} \qquad (5.26)$$

As the result, we get

$\ln \Xi = \ln \Xi_0$ + Sum of all the connected diagrams composed of solid lines ——▸—— dotted lines. . . . . The factors to be associated to ——▸—— and . . . . are

$$\frac{1}{\zeta_l - \epsilon_k} \quad \text{and } v_q \qquad (5.27)$$

The sum of ζ_l's and k's are to be taken to

conserve at any vertex. E. g. for a part of diagram shown in Fig. 5-4,

$$k_1' - k_1 = k_2' - k_2 = q.$$

$$\zeta_1' - \zeta_1 = \zeta_2' - \zeta_2.$$

It is interesting to see that the sum of ζ_1 variables in addition to the sum of momentum conserves in interaction.

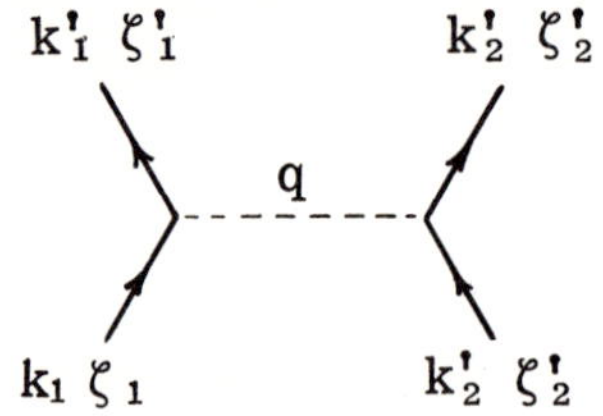

Statistical Mechanics of Irreversibility

Robert W. Zwanzig

LECTURE I

Part 1

The purpose of these lectures is to describe and explain some mathematical techniques that are useful in quantum statistical mechanics. These techniques, not yet very well known, are connected with the use of Liouville operators and projection operators. Their main illustration here will be in the derivation of certain "master equations" of interest in the theory of irreversible processes.

First, a quick survey of some background material will be useful. This is connected with the following question.

The density matrix (operator) at time t is $\hat{\rho}(t)$. Its evolution is determined by the Hamiltonian operator

$$\hat{H} = H_0 + \lambda \hat{V}. \tag{1}$$

where $\hat{H}_0$ is some unperturbed Hamiltonian and $\lambda\hat{V}$ is a perturbation. The strength of the perturbation is measured by λ.

The unperturbed system $\hat{H}_0$ may consist of non-interacting particles, quasi-particles, etc.; it should have the property that we can find its eigenvalues and eigenstates,

$$\hat{H}_0 | n\rangle = E_n | n\rangle \tag{2}$$

We shall work in this unperturbed representation. The density matrix is

$$\rho_{mn}(t) = \langle m | \hat{\rho}(t) | n \rangle \tag{3}$$

and the Hamiltonian is

$$H_{mn} = E_m \delta_{mn} + \lambda V_{mn} \tag{4}$$

For simplicity we assume that E_m contains all diagonal contributions, so that

$$V_{mm} = 0 \tag{5}$$

The diagonal elements of $\hat{\rho}$ describe the occupation of unperturbed states,

$$\rho_{mm}(t) = \text{prob. of observing the system in state } |m\rangle \text{ at time t.} \tag{6}$$

In the absence of a perturbation, $\rho_{mm}(t)$ is constant. This can be seen from the equation of motion,

$$\frac{\partial \hat{\rho}}{\partial t} = -i[\hat{H}, \hat{\rho}] \tag{7}$$

$$\frac{\partial \rho_{mn}}{\partial t} = -i \sum_l (H_{ml} \rho_{ln} - \rho_{ml} H_{ln})$$

and, for the diagonal elements, on putting in $E_m \delta_{mn} + \lambda V_{mn}$,

$$\frac{\partial \rho_{mm}}{\partial t} = -i \sum_l (\lambda V_{ml} \rho_{lm} - \rho_{ml} \lambda V_{lm}) \tag{8}$$

So if no perturbation, ρ_{mm} is constant in time.

With this long prelude, we can now state the problem of interest: What is the time dependence of the occupation probability $\rho_{mm}(t)$ as a result of the perturbation λV_{mn}?

It turns out that there are at least two solutions to this problem. The first one is obtained from the Liouville equation just written,

$$\frac{\partial \hat{\rho}}{\partial t} = -i[\hat{H}, \hat{\rho}] \tag{9}$$

The formal solution to this equation is as follows,

$$\hat{\rho}(t) = e^{-it\hat{H}}\hat{\rho}(0)\, e^{+it\hat{H}} \tag{10}$$

so that

$$\rho_{mm}(t) = \sum_j \sum_k (e^{-itH})_{mj}\, \rho_{jk}(0)\, (e^{+it\hat{H}})_{km} \tag{11}$$

This determines in principle and exactly the desired time dependence, in terms of the initial values.

The most common special case is where the density matrix is initially diagonal,

$$\rho_{mn}(0) = \rho_{mm}(0)\, \delta_{mn} \tag{12}$$

Then

$$\rho_{mm}(t) = \sum_n \left| (e^{-itH})_{mn} \right|^2 \rho_{nn}(0) \tag{13}$$

so that

$$|(e^{-it\hat{H}})_{mn}|^2$$

gives the probability of transition from a state n at $t = 0$ to a state m at t.

We shall return to this solution in a little while. First we look at the other solution. This was given by Pauli:

$$\frac{\partial \rho_{mm}(t)}{\partial t} \cong \sum_n \left[W_{mn}\, \rho_{nn}(t) - W_{nm}\rho_{mm}(t) \right] \tag{14}$$

The coefficients W_{mn} are transition rates, given in the limit of weak interaction by the "Golden Rule,"

$$W_{mn} = 2\pi\lambda^2\, |V_{mn}|^2\, \delta(E_m - E_n) \tag{15}$$

Without actually solving Pauli's master equation, we can see that its solutions are not identical with the exact solutions. To do this we calculate the time derivatives at $t = 0$, with the special initial condition

$$\left\{ \begin{array}{l} \rho_{11}(0) = 1 \\ \text{all other } \rho_{mn}(0) = 0 \end{array} \right\} \tag{16}$$

Then the exact equation (8) is

$$\left(\frac{\partial \rho_{11}}{\partial t}\right)_0 = -i \sum_k \left[\lambda V_{1k}\, \rho_{1k}(0) - \lambda \rho_{1k}(0)\, V_{k1} \right] \tag{17}$$

But V_{1k} vanishes for $k = 1$ by the way we originally defined the perturbation, and $\rho_{1k}(0)$ vanishes for $k \neq 1$. Therefore the whole RHS vanishes, or

$$\left(\frac{\partial \rho_{11}}{\partial t}\right)_0 = 0 \tag{18}$$

But we can calculate this also from the master equation,

$$\left(\frac{\partial \rho_{11}}{\partial t}\right)_0 \cong \sum_n W_{1n}\,\rho_{nn}(0) - \sum_n W_{n1}\,\rho_{11}(0) \tag{19}$$

Note that W_{1n} vanishes for $n = 1$ and $\rho_{nn}(0)$ vanishes for $n \neq 1$. Therefore

$$\left(\frac{\partial \rho_{11}}{\partial t}\right)_0 \cong -\left(\sum_n W_{n1}\right)\rho_{11}(0) \tag{20}$$

The RHS cannot vanish because it contains a sum $\sum_n W_{n1}$ of intrinsically positive quantities. Se we have a contradiction.

In spite of this, the Pauli master equation is often nearly correct. We want to understand how this happens, and what the limitations are.

THE STANDARD DERIVATION

To begin with, let's look at the customary derivation of Pauli's equation. It takes the form of a gain-loss equation

$$\begin{aligned} \Delta\rho_{mm}(t) &= \text{net increase of occupation of state } |m\rangle \text{ during initial } \Delta t \\ &= \text{gain in } |m\rangle - \text{loss from } |m\rangle \end{aligned} \tag{21}$$

The gain can come only from other states. This is "obviously"

$$\begin{aligned} \text{gain in } |m\rangle &= \sum_n{}' \,[\text{prob. of transition from n to m during } \Delta t] \\ &\qquad \times [\text{occupation of } |n\rangle\,] \\ &= \sum_n{}' \,[W_{mn}\,\Delta t] \times \rho_{nn}(t) \end{aligned} \tag{22}$$

and the loss can come only from transitions to other states,

$$\text{loss from } |m\rangle = \left(\sum_n{}' W_{nm}\,\Delta t\right) \times \rho_{mm}(t) \tag{23}$$

Now we write

$$\frac{\partial \rho}{\partial t} = \frac{\Delta \rho}{\Delta t} \tag{24}$$

and we get the master equation. To conclude the derivation we must also get the "Golden Rule."

The derivation given by Schiff is typical of those in text-books. It is based on first order perturbation theory. We start with Equation (15), where

$$|(e^{-it\hat{H}})_{mn}|^2 = \text{prob. of transition from } |n\rangle \text{ initially to } |m\rangle \text{ at time t.} \tag{25}$$

Use the perturbation expansion

$$e^{-it\hat{H}} = e^{-it\hat{H}_0} - i\int_0^t dt_1\, e^{-i(t-t_1)\hat{H}_0}\, \lambda\hat{V}\, e^{-it_1\hat{H}_0} \quad \text{higher orders in } \lambda. \tag{26}$$

As the general expansion theorem, note that

$$e^{-it\hat{H}} = e^{-it\hat{H}_0} - i\int_0^t dt_1\, e^{-i(t-t_1)\hat{H}_0}\, \lambda\hat{V}\, e^{-it_1\hat{H}} \tag{27}$$

This can be verified by differentiating. Eq. (26) is obtained by substituting

$\exp(-it_1\hat{H}_0)$ for $\exp(-it_1\hat{H})$ on the RHS.

Now we take the (m, n) matrix element and do the time integration. We restrict the discussion to transitions between different states, or $m \neq n$. Then

$$(e^{-it\hat{H}_0})_{mn} = e^{-itE_m}\delta_{mn} = 0 \tag{28}$$

and

$$\begin{aligned}(e^{-it\hat{H}})_{mn} &= i\int_0^t dt_1\, e^{-i(t-t_1)E_m}\,(\lambda V)_{mn}\, e^{-it_1E_n} \\ &= -\lambda V_{mn}\,\frac{e^{it\omega_{mn}}-1}{\omega_{mn}} \times e^{-itE_m}\end{aligned} \tag{29}$$

when

$$\omega_{mn} = E_m - E_n \tag{30}$$

We take the square; after some algebra, we get

$$|(e^{-it\hat{H}})_{mn}|^2 = 4\lambda^2 \, |V_{mn}|^2 \left[\frac{\sin^2 \frac{1}{2} \omega_{mn} t}{\omega_{mn}^2} \right] \tag{31}$$

Let's call the quantity in square brackets

$$\Delta(\omega, t) = \frac{\sin^2 \frac{1}{2} \omega t}{\omega^2} \tag{32}$$

Its shape is shown in Fig. 1

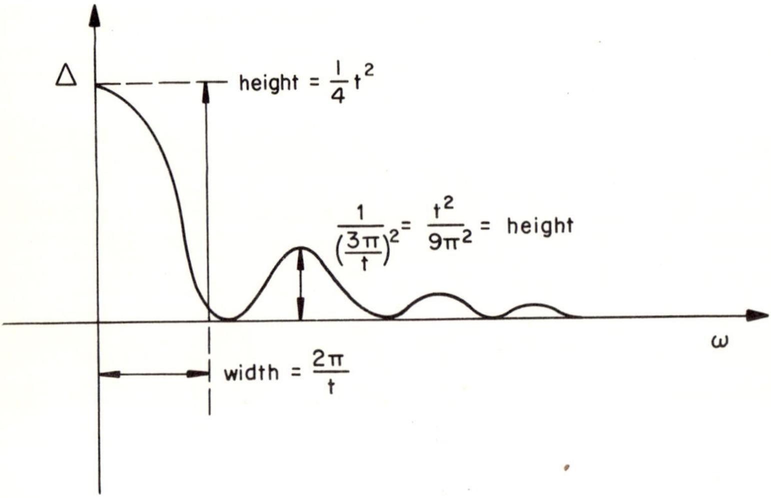

Figure 1

$$\text{area} \quad \sim \frac{1}{4} t^2 \times \frac{2\pi}{t} \to \frac{\pi}{2} t$$

Actually,

$$\lim_{t \to \infty} \Delta(\omega, t) = \frac{\pi}{2} \, t \, \delta(\omega) \tag{33}$$

So the transition probability from n to m in t is

$$4\lambda^2 |V_{mn}|^2 \ \Delta(\omega, t) + 0(\lambda^3) \tag{34}$$

and as t gets very long this approaches

$$\rightarrow 4\lambda^2 |V_{mn}|^2 \frac{\pi}{2} t\, \delta(\omega_{mn}) \tag{35}$$

So, for large t we get for the transition probability during t,

$$t W_{mn} \tag{36}$$

with our earlier "Golden Rule" definition of W_{mn}.

But now we have to understand how to interpret the limit.

(1) t is never actually infinite, so the delta function is never exact. However, the area is always exactly proportioned to t.

(2) The <u>true</u> energy eigenvalues are of the order of magnitude.

$$E_{time,\ m} = E_{unperturbed,\ m} + 0\ (\lambda^2)$$

(This comes from perturbation theory). "Because true total energy is conserved" we expect that

$$E_m + 0(\lambda^2) = E_n + 0(\lambda^2)$$

or

$$\omega_{mn} = 0(\lambda^2) \tag{37}$$

for the physically possible transitions. This means that, for small λ,

$$\omega_{mn}\, t = 0(\lambda^2 t) \tag{38}$$

But for ω to be inside the first peak of the function $\Delta(\omega, t)$, we require

$$|\omega t| \lesssim 0(2\pi)$$

or, for typical values of ω,

$$0(\lambda^2 t) \lesssim 0(2\pi) = 0(1) \tag{39}$$

So, by picking $\lambda^2 t$ small enough, even with λ small and t large, we are usually within the main peak of $\Delta(\omega, t)$.

(3) A third point in the standard derivation is to suppose that the unperturbed spectrum is quite dense. For example, in a many body system with N particles, the number of states within any internal $d\omega$ is of the order

$$N\rho(\omega)\,d\omega \tag{40}$$

where $\rho(\omega)$ is a density of states, of order (1) as $N \to \infty$. So, for large N, there will be many unperturbed states within the range

$$\omega_{mn} = 0(\lambda^2) \tag{41}$$

These all have roughly the same unperturbed energy. Let's now sum over these (final) states

$$\sum_{\omega_{mn} = 0(\lambda^2)} |(e^{-it\hat{H}})_{mn}|^2$$

$$= \sum_{\omega_{mn}=0(\lambda^2)} 4\lambda^2 |V_{mn}|^2 \Delta(\omega_{mn}, t) \tag{42}$$

$$\to \int d\omega_{mn} N\rho(\omega_m) 4\lambda^2 |V_{nm}|^2 \Delta(\omega_{mn}, t)$$

Now, suppose that $\rho(\omega)$ and V_{nm} are "smooth" functions of ω_{mn} over the range

$$\omega_{mn} = 0(\lambda^2)$$

Then, for large t we can replace the integral by

$$N\rho(\omega_m)\, 4\lambda^2 |V_{nm}|^2 \int d\omega_{mn}\, \Delta(\omega_{mn}, t) \tag{43}$$

But the latter integral is known, exactly, so we get

$$\to \underbrace{N\rho(\omega_m)\, 2\pi\lambda^2 |V_{nm}|^2}_{W_{mn}} \times t \tag{44}$$

And we have here just another form of W_{mn}

In deriving the expression we used the following order of magnitude estimates

(1) λ small
(2) t large
(3) $\lambda^2 t < 0(2\pi)$
(4) $\omega_{mn} = 0(\lambda^2)$
(5) $\rho(\omega) = 0(1)$ as $N \to \infty$

Note especially the third estimate. This will come up again in Van Hove's derivation.

Having found a transition probability proportioned to t, we get a time independent transition rate,

$$\text{rate} = \frac{d}{dt}(\text{prob.}) \tag{45}$$

This concludes our survey of the old-fashioned derivations of Pauli's master equations.

LECTURE I

Part 2

VAN HOVE'S CONTRIBUTIONS

Van Hove made two principal contributions to the modern theory of the master equation. His first (1955) was a more careful and rigorous derivation of Pauli's equation, for small λ only. His later contribution (1957) and (1959) was a generalization to higher orders in λ. Let's consider the first now.

We need the full perturbation expansion of (26), obtained by iterating (27),

$$e^{-it\hat{H}} = e^{-it\hat{H}_0} - i\int_0^t dt_1\, e^{-i(t-t_1)\hat{H}_0}\,\lambda\hat{V}e^{-it_1\hat{H}_0}$$
$$+ (-i)^2\int_0^t dt_1\int_0^t dt_2\, e^{-i(t-t_1)\hat{H}_0}\,\lambda\hat{V}e^{-i(t_1-t_2)\hat{H}_0}\,\lambda\hat{V}e^{-it_2\hat{H}_0} \tag{46}$$
$$+0(\lambda^3)$$

And, we need the square of the expansion, or

$$|(e^{-it\hat{H}})_{mn}|^2 = (e^{it\hat{H}})_{nm}(e^{-it\hat{H}})_{mn} \tag{47}$$

Any term in the expansion of the product can be indicated by a graph of the following sort,

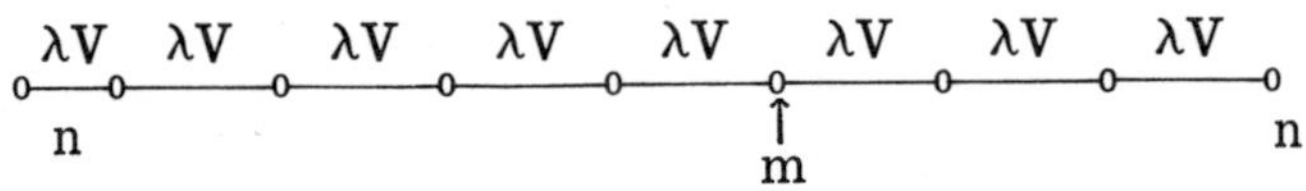

This particular one shows 5th order from the left and 3rd order from the right. The graph needed in the original Pauli theory is just

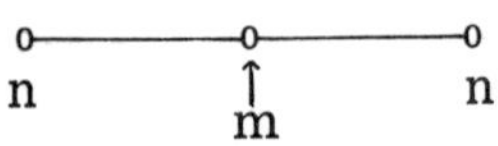

To the next order we get (for $n \neq m$)

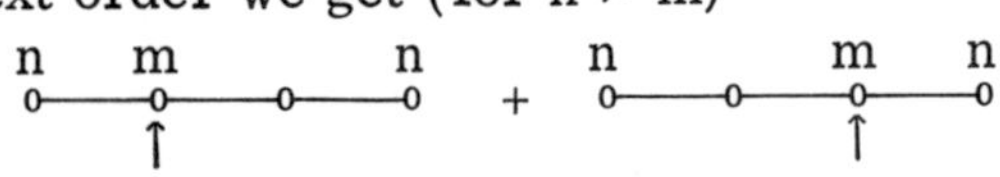

and in the fourth order,

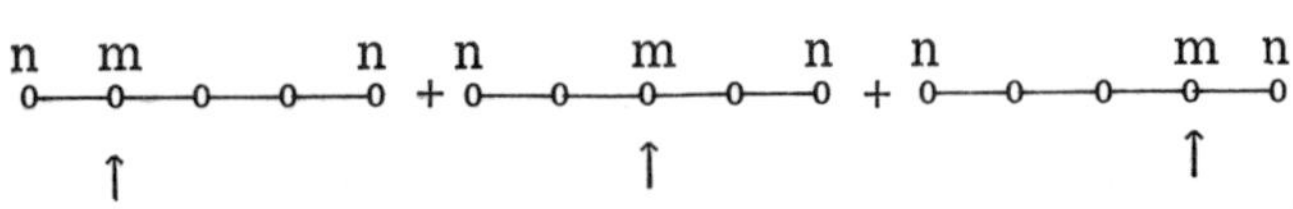

Now Van Hove observed that, for reasonable interactions of of the sort that we have had in mind all along, whenever one has a combination like

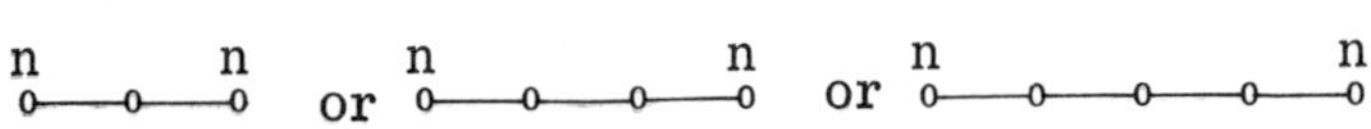

where all the interior indices are different from each other and different from n, then the integral gives asymptotically a factor t.

The reason for this is just the same as we saw in the earlier discussion: one gets a function of time that looks asymptotically like $t\,\delta(\omega)$, where the δ function is not really infinitely sharp, but only nearly so.

Note that the second order term has the character:

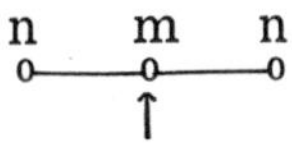

$\sim \lambda^2 t$

In third order we can have only

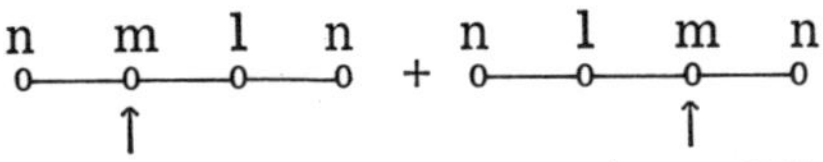

$\sim \lambda^3 t$

In fourth order, however, we get the possibility of

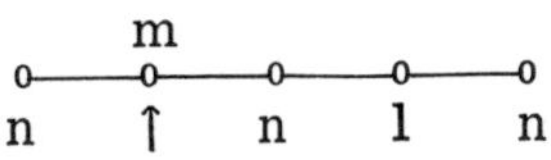

$\sim (\lambda^2 t)^2$

In sixth order we can get

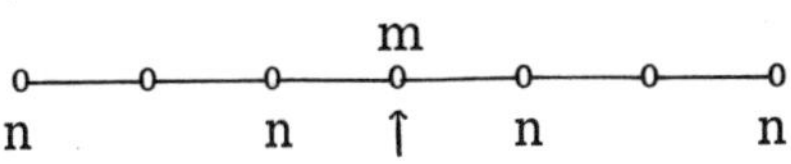

$\sim (\lambda^2 t)^3$

And so on.
The important point here is that higher powers of t will occur, in fact the whole series looks asymptotically like

$$\begin{aligned} |\ \ |^2 &= (\ \ )\,\lambda^2 t + (\ \ )\,\lambda^3 t + (\ \ )^4 \lambda\ t + \dots \\ &+ (\ \ )(\lambda^2 t)^2 + (\ \ )(\lambda^3 t)^2 + \dots \\ &+ (\ \ )(\lambda^3 t)^3 + (\ \ )(\lambda^3 t)^3 + \dots \end{aligned} \tag{48}$$

Van Hove noticed that the limit

$$\left.\begin{array}{l} \lambda \to 0 \\ t \to \infty \end{array}\right\} \quad \lambda^2 t = \text{constant} \tag{49}$$

eliminated all but the set of powers of $\lambda^2 t$. In the limit, which is just like the one we used earlier,

$$|\ \ |^2 \to (\ \ )\lambda^2 t + (\ \ )(\lambda^2 t)^2 + (\ \ )(\lambda^2 t)^3 \dots \tag{50}$$

He then observed that upon differentiation,

$$\frac{d}{dt}|(e^{-itH})_{mn}|^2 = \sum_{l}{}' W_{nl}|(e^{-it\hat{H}})_{lm}|^2 - \sum_{l}{}' W_{ln}|(e^{-it\hat{H}})_{nm}|^2 \tag{51}$$

so that if we start out in the particular state m, and

$$\rho_{mn}(t) = |(e^{-it\hat{H}})_{nm}|^2 \tag{52}$$

we get

$$\frac{d\rho_{nn}}{dt} = \sum_{l}(W_{nl}\rho_{ll} - W_{ll}\rho_{nn}) \tag{53}$$

But because of superpostion of initial conditions, the result, not containing m, is in fact independent of the choice of initial state.

REFERENCES

W. Pauli, Festschrift zum 60. Geburtstag A. Sommerfeld (Hirzel, Leipzig), (1928).
L. Van Hove, Physica, **21** 517, (1955), Physica, **23** 441, (1957), Physica, **25** 268, (1959), "Fundamental Problems in Statistical Mechanics" (ed. by E. G. D. Cohen, North-Holland, pp. 137-172.) (1962).
A. Janner, Helv. Phys. Acta, **35**, 47 (1962).
R. J. Swenson, J. Math. Phys. **3**, 1017 (1962).
R. Zwanzig, J. Chem. Phys. **33**, 1338, (1960).
E. W. Montroll, "Fundamental Problems in Statistical Mechanics," pp. 230-249 (1962).
H. C. Praddaude, Ann. Phys. **22**, 210 (1963).
R. Zwanzig, "Lectures in Theoretical Physics III." pp. 106-141, (Interscience Publ., N. Y.) (1961).

LECTURE II

Part 1

The second lecture will be devoted to Liouville operator techniques in quantum statistical mechanics.

First, we make a table of the standard notations. The "Liouville equation" is

$$\left\{ \begin{aligned} \frac{\partial \hat{\rho}(t)}{\partial t} &= -i[\hat{H}, \hat{\rho}(t)] \\ \frac{\partial \rho_{mn}(t)}{\partial t} &= -i\sum_{l} (H_{ml}\rho_{ln}(t) - \rho_{ml}(t)\, H_{ln}) \end{aligned} \right\}$$

The formal solution is

$$\left\{ \begin{aligned} \hat{\rho}(t) &= e^{-it\hat{H}}\, \hat{\rho}(0)\, e^{+it\hat{H}} \\ \rho_{mn}(t) &= \sum_{j}\sum_{k} (e^{-it\hat{H}})_{mj}\rho_{jk}(0)\, (e^{+it\hat{H}})_{kn} \end{aligned} \right.$$

Resolvents and Laplace transforms: sometimes one wants to work with the Laplace transform of the density matrix.

Define

$$\hat{g}(s) = \int_{0}^{\infty} dt\, e^{-st}\hat{\rho}(t)$$

The transform of the equation of motion of $\hat{\rho}$ is

$$s\,\hat{g}(s) - \hat{\rho}(0) = -i[\hat{H}, \hat{g}(s)]$$

$$sg_{mn}(s) - \rho_{mn}(0) = -i\sum_{l}[H_{ml}g_{ln}(s) - g_{ml}(s)H_{ln}]$$

The latter are linear equations for $g_{ml}(s)$ which can in principle be solved for g. The Laplace transform can be written by means of resolvents also. Let

$$\hat{R}(z) = \frac{1}{\hat{H} - z\hat{1}}$$

$$R_{mn}(z) = [(\hat{H} - z\hat{1})^{-1}]_{mn}$$

Note that the singularities of R come at the (real) eigenvalues of $\hat{H}$. So,

$$e^{\pm it\hat{H}} = \frac{1}{2\pi i}\oint dz\,\frac{e^{\pm itz}}{z - \hat{H}}$$

$$= -\frac{1}{2\pi i}\oint dz\,e^{\pm itz}\hat{R}(z)$$

where the contour encloses the part of the real axis occupied by the exact eigenvalues of $\hat{H}$.
Then

$$\hat{g}(s) = \int_0^\infty dt\,e^{-st}\,e^{-it\hat{H}}\hat{\rho}(0)\cdot\frac{-1}{2\pi i}\oint dz\,e^{itz}\,\hat{R}(z)$$

$$= \frac{-1}{2\pi i}\oint dz\int_0^\infty dt\,e^{-st}\,e^{izt}\,e^{-i\hat{H}t}\hat{\rho}(0)\,\hat{R}(z)$$

$$= \frac{-1}{2\pi i}\oint dz\,\frac{1}{s - iz + i\hat{H}}\,\hat{\rho}(0)\,\hat{R}(z)$$

$$= -\frac{1}{2\pi i}\oint dz . - i\,\hat{R}(z - is)\hat{\rho}(0)\,\hat{R}(z)$$

where the integration has to be taken around the poles of

$\hat{R}(z)$ only. So

$$\hat{g}(s) = \frac{1}{2\pi} \oint dz\, \hat{R}(z - is)\, \hat{\rho}(0)\, \hat{R}(z)$$

gives the Laplace transform of the density matrix in terms of Resolvents.

This equation is equivalent to the starting point of Van Hove's general derivation of the master equation. The only difference is that he uses Fourier transforms instead of Laplace transforms.

The inversion of a Laplace transform is

$$\hat{\rho}(t) = \frac{1}{2\pi i} \int_{\epsilon - i\infty}^{\epsilon + i\infty} ds\, e^{st}\, \hat{g}(s)$$

with a contour to the right of all singularities of $\hat{g}(s)$.

Van Hove does a perturbation expansion on the two resolvents and then collects terms and inverts the transform.

Now we want to rewrite all these formulas in the "Liouville Operator" notation.

$$\frac{\partial \hat{\rho}}{\partial t} = -i\, [\hat{H}, \hat{\rho}(t)] = -i\, L\, \hat{\rho}(t)$$

where (for arbitrary operator $\hat{A}$)

$$\hat{B} = L\hat{A} = [\hat{H}, \hat{A}]$$

This is a linear operator that turns an operator $\hat{A}$ into a new operator $\hat{B}$.

With subscripts,

$$\frac{\partial \rho_{mn}}{\partial t} = -i \sum_{m'} \sum_{n'} L_{mn\, m'n'}\, \rho_{m'n'}$$

$$L_{mn\, m'n'} = H_{mm'}\, \delta_{nn'} - H_{n'n}\, \delta_{mm'}$$

$$= H_{mm'}\, \delta_{nn'} - \delta_{mm'}\, (H^{*})_{nn'}$$

The operator L can be regarded as a <u>tetradic</u> operator (four subscripts). The multiplication rule for tetradics is

$$(L_1 L_2)_{mnm'n'} = \sum_k \sum_l (L_1)_{mnkl} (L_2)_{klm'n'}$$

The identity tetradic is

$$(1)_{mnm'n'} = \delta_{mm'} \delta_{nn'}$$

so that

$$(1 \cdot \hat{A})_{mn} = \sum_{m'} \sum_{n'} (1)_{mnm'n'} A_{m'n'} = A_{mn}$$

and

$$(1 \cdot L)_{mnm'n'} = L_{mnm'n}$$

Evidently tetradics behave much like matrices. In fact, one can reduce their algebra to that of ordinary matrices by the following trick.

We start by representing operators by matrices

$$\hat{A} \rightarrow A_{mn}$$

Next, we pick some ordering sequence for the subscripts, for example

$$\alpha = 1 \leftrightarrow (m, n) = (11)$$

$$\alpha = 2 \leftrightarrow (m, n) = (12)$$

$$\alpha = 3 \leftrightarrow (m, n) = (21)$$

$$\alpha = 4 \leftrightarrow (m, n) = (13)$$

$$\alpha = 5 \leftrightarrow (m, n) = (22)$$

$$\alpha = 6 \leftrightarrow (m, n) = (31)$$

$$\alpha = 7 \leftrightarrow (m, n) = (14)$$

$$\alpha = 8 \leftrightarrow (m, n) = (23)$$

$$\alpha = 9 \leftrightarrow (m, n) = (32)$$

$$\alpha = 10 \leftrightarrow (m, n) = (41)$$

etc.

So we identify

$$\hat{A} \rightleftarrows A_{mn} \to A_{(\alpha)} \to \text{"vector } \vec{A}\text{"}$$

Then the formula

$$\hat{B} = L\hat{A}$$

can be written as

$$B_{(\alpha)} = \sum_{\beta=1}^{\infty} L_{(\alpha)(\beta)} A_{(\beta)} \qquad (\alpha = 1, 2, 3, \ldots)$$

so that $L_{(\alpha)(\beta)}$ behaves like a matrix of two subscripts.

This means that (with due caution!) the algebra of L can be reduced to the algebra of matrices.

Now let's return to Liouville equation,

$$\frac{\partial\hat{\rho}}{\partial t} = i L\hat{\rho}(t)$$

This has the solution

$$\hat{\rho}(t) = e^{-itL}\, \hat{\rho}(0)$$

where

e^{-itL} is a tetradic operator,

$$e^{-itL} = \sum_{0}^{\infty} \frac{(-it)^n}{n!} L^n$$

Because we have defined the multiplication of L, this is a good quantity. The component form is

$$\rho_{mn}(t) = \sum_{m'}\sum_{n'} (e^{-itL})_{mnm'n'}\, \rho_{m'n'}(0)$$

These are to be compared with the traditional forms

$$\hat{\rho}(t) = e^{-it\hat{H}}\hat{\rho}(0)\, e^{+it\hat{H}}$$

$$\rho_{mn}(t) = \sum_{m'}\sum_{n'} (e^{-it\hat{H}})_{mm'}\, \rho_{m'n'}(0)\, (e^{it\hat{H}})_{nn'}$$

so we see that

$$(e^{-itL})_{mnm'n'} = (e^{-it\hat{H}})_{mm'}\, (e^{it\hat{H}})_{nn'}$$

Let's verify the first power of t just for fun.

$$
\begin{aligned}
(\text{LHS}) &= (1)_{mnm'n'} - it\,L_{mnm'n'} + \ldots \\
&= \delta_{mm'}\,\delta_{nn'} - it\,H_{mm'}\,\delta_{nn'} \\
&+ it\,\delta_{mm'}\,H_{n'n} + 0\,(t^2)
\end{aligned}
$$

Now do

$$
\begin{aligned}
\text{RHS} &= (\hat{1} - it\hat{H})_{mm'}\,(1 + itH)_{n'n} \\
&= (\delta_{mm'} - it\,H_{mm'} + \ldots)(\delta_{nn'} + it\,H_{n'n} + \ldots) \\
&= \delta_{mm'}\,\delta_{nn'} - it\,H_{mm'}\,\delta_{nn'} + it\,\delta_{mm'}\,H_{n'n} + \ldots
\end{aligned}
$$

So RHS = LHS to order t. Obviously this must be true to all orders.

To go further, we can do perturbation expansions with L

if $\hat{H} = \hat{H}_0 + \hat{H}_1$ then clearly

$$L = L_0 + L_1$$

The operator e^{-itL} obeys the integral equation

$$e^{-itL} = e^{-itL_0} - i\int_0^t dt_1\, e^{-i(t-t_1)L_0}\, L_1 e^{-itL}$$

which gives, on differentiation,

$$\frac{d}{dt}\, e^{-itL} = -i\,L\,e^{-itL}$$

as it should. We can use the integral equation by iteration to get

$$
\begin{aligned}
e^{-itL} &= e^{-itL_0} - i\int_0^t dt_1\, e^{-i(t-t_1)L_0}\, L_1 e^{-it_1L_0} \\
&+ (-i)^2 \int dt_1 \int dt_2\, e^{-i(t-t_2)L_0}\, L_1{}^{-i(t_1-t_2)L_0}\, L_1 e^{-it_2L_0}
\end{aligned}
$$

etc.

This is all just as one would do with Hamiltonian operators. It has the advantage, however, that the perturbation comes always in one place. Compare this with the expansion of the standard form

$$e^{-itH}\hat{\rho}(0)\,e^{+it\hat{H}}$$

where one gets $\hat{H}_1$ on both sides of $\hat{\rho}(0)$.

But the real advantage appears in the resolvent form. From

$$\hat{g}(s) = \int_0^\infty dt\, e^{-st}\,\hat{\rho}(t)$$

we get

$$s\hat{g}(s) - \hat{\rho}(0) = -iL\,\hat{g}(s)$$

"Solve" this for $\hat{g}$,

$$\hat{g}(s) = \frac{1}{s+iL}\,\hat{\rho}(0)$$

This is very much neater than the earlier

$$\hat{g}(s) = \frac{1}{2\pi}\oint dz\,\hat{R}(z-is)\,\hat{\rho}(0)\,\hat{R}(z)$$

Note also that this connects with the formal solution:

$$g\quad \hat{g}(s) = \int_0^\infty dt\, e^{-st}e^{-itL}\hat{\rho}(0) = \frac{1}{s+iL}\,\hat{\rho}(0)$$

and we get this also by inverting the Laplace transform. Finally, observe the identity

$$\frac{1}{s+i(L_0+L_1)} = \frac{1}{s+iL_0} - i\,\frac{1}{s+iL_0}\,L_1\,\frac{1}{s+i(L_0+L_1)}$$

which iterates to the perturbation expansion

$$= \frac{1}{s+iL_0} - i\,\frac{1}{s+iL_0}\,L_1\,\frac{1}{s+iL_0}$$

$$+ (i)^2\,\frac{1}{s+iL_0}\,L_1\,\frac{1}{s+iL_0}\,L_1\,\frac{1}{s+iL_0} - 0(L_1^3)$$

So much for the generalities of notation.

LECTURE II

Part 2

Next, we consider the derivation of the master equation.

What we want is the diagonal part of the density matrix. Let's introduce a projection operator D to select the diagonial part.

$$(D\hat{\rho})_{mn} \equiv \rho_{mm}\,\delta_{mn}$$

This operation can be represented explicitly by a tetradic,

$$D_{mnm'n'} = \delta_{mm'}\,\delta_{nn'}\,\delta_{mn}$$

which evidently has the desired character. Similarly, the operator that selects the non-diagonal part is 1 - D, where

$$(1 - D)_{mnm'n'} = \delta_{mm'}\,\delta_{nn'}\,(1 - \delta_{mn}),$$

Now we use D and 1 - D to separate the Liouville equation into two parts,

$$D s \hat{g}(s) - D\hat{\rho}(0) = -iDL\hat{g}(s)$$

$$(1 - D)\; s\hat{g}(s) - (1 - D)\hat{\rho}(0) = -i(1 - D)\; L\hat{g}(s)$$

For convenience in notation we write

$$\hat{g}(s) = \hat{g}_1(s) + \hat{g}_2(s)$$

$$\hat{g}_1(s) = D\hat{g}(s)$$

$$\hat{g}_2(s) = (1 - D)\,\hat{g}(s)$$

Then the two parts of Liouville equation contain $\hat{g}_1$ and $\hat{g}_2$,

$$s\hat{g}_1(s) - \hat{\rho}_1(0) = -iDL\hat{g}_1(s) - iDL\hat{g}_2(s)$$

$$s\hat{g}_2(s) - \hat{\rho}_2(0) = -i(1 - D)\,\hat{g}_1(s)\,L - i(1 - D)\,L\hat{g}_2(s)$$

Solve the second equation for $\hat{g}_2$:

$$[s + i(1 - D)L]\hat{g}_2 = \hat{\rho}_2(0) - i(1 - D)L\hat{g}_1$$

$$\hat{g}_2 = \frac{1}{s + i(1 - D)L}\hat{\rho}_2(0) - \frac{1}{s + i(1 - D)L}(1 - D)L\hat{g}_1$$

Put this back in the first equation,

$$s\hat{g}_1(s) + iDL\hat{g}_1(s) = \hat{\rho}_1(0)$$

$$- iDL\frac{1}{s + i(1 - D)L}\hat{\rho}_2(0)$$

$$- DL\frac{1}{s + i(1 - D)L}(1 - D)L\hat{g}_1(s)$$

Note that the equation relates $\hat{g}_1(s)$ to itself and to <u>both initial parts.</u>

Next we invert the Laplace transform. The only tricky point here is that the inverse transform of a product is a convolution,

$$\int_0^\infty dte^{-st}\int_0^t dt_1\ a(t_1)b(t - t_1)$$

$$= \int_0^\infty dte^{-st}\ a(t) \cdot \int_0^\infty dt'\ e^{-st'}b(t')$$

So the equation for $\dot{g}_1$ gives

$$\frac{d\hat{\rho}_1(t)}{dt} = - iDL\hat{\rho}_1(t) - iDLe^{-i(1-D)Lt}\ \hat{\rho}_2(0)$$

$$-\int_0^t dt_1\left[DLe^{-it_1(1-D)L}\ (1 - D)L\right]\hat{\rho}_1(t - t_1)$$

The exciting thing here is that when the non-diagonal part vanishes initially,

$$\hat{\rho}_2(0) = 0$$

or the density matrix is initially diagonal, then we have an equation that contains $\hat{\rho}_1$ only. Even though the non-diagonal part will generally not vanish after t = 0 we don't have to know what it is.

The quantity

$$g(t) = e^{-i(1-D)Lt}$$

can be interpreted two ways:

(1) As the solution of

$$\frac{dg}{dt} = -i(1 - D)Lg$$

and (2) by the expansion

$$g = 1 - i(1 - D)LD + \frac{i^2}{2}(1 - D)L(1 - D)Lt^2 \ldots$$

The general master equation is therefore

$$\frac{d\hat{\rho}_1(t)}{dt} = -iDL\hat{\rho}_1(t) - \int_0^t dt_1 K(t_1)\hat{\rho}_1(t - t_1)$$

when $\hat{\rho}_2(0) = 0$ and

$$K(t_1) = DLe^{-i(1-D)Lt_1}(1 - D)L$$

Now we take components

$$\langle n|\hat{\rho}_1(t)|n\rangle = \rho_{nn}(t)$$

and

$$\frac{d\rho_{nn}(t)}{dt} = -i[L\hat{\rho}(t)]_{nn} - \int_0^t dt_1 [K(t_1)\hat{\rho}_1(t - t_1)]_{nn}$$

The first part of the RHS vanishes. This is how;

$$[L\hat{\rho}_1]_{nn} = \sum_a \sum_b L_{nnab}(\hat{\rho}_1)_{ab} = \sum_a L_{nnaa}\rho_{aa}$$

But

$$L_{mnaa} = H_{na}\,\delta_{na} - \delta_{na}\,H_{an}$$
$$= (H_{aa} - H_{aa})\,\delta_{na} = 0$$

This gives:

$$\frac{d\rho_{nn}(t)}{dt} = -\int_0^t dt_1\,[K(t_1)\cdot\hat{\rho}_1(t - t_1)]_{nn}.$$

To finish this we write

$$[K\cdot\hat{\rho}_1]_{nn} = \sum_c \sum_d K_{nncd}\,(\hat{\rho}_1)_{cd}$$
$$= \sum_m K_{nnmm}\rho_{mm}$$

Thus,

$$\frac{d\rho_{nn}(t)}{dt} = -\sum_m \int_0^t dt_1\,K_{nnmm}\rho_{mm}$$

This is beginning to look like a master equation.
Lets work out K in detail, step by step

$$K_{nnmm}(t) = [Le^{-it_1(1-D)L}\quad(1 - D)L]_{nnmm}$$

First, note that

$$[LX]_{nnmm} = \Sigma\Sigma L_{nncd}\,X_{cdmm}$$

$$=\Sigma\Sigma\,L^0_{nncd}\,X_{cdmm} + \Sigma\Sigma\,L^1_{nncd}\,X_{cdmm}$$

look at the L^0 part,

$$\Sigma\Sigma(H^0_{nc}\,\delta_{nd} - \delta_{nc}\,H^0_{dn})X_{cdmm}$$

$$= \Sigma\Sigma(E_n\,\delta_{nc}\,\delta_{nd} - E_n\,\delta_{nc}\,\delta_{nd})X_{cdmm} = 0$$

So only L_1 appears, and

$$K_{nnmm} = [L_1\,e^{-it_1(1-D)L}\quad(1 - D)L]_{nnmm}$$

Coming in from the other end, we find

$$[(1 - D)L]_{abmm} = (1 - \delta_{ab})L_{abmm}$$
$$= (1 - \delta_{ab})(L^0_{abmm} + L^1_{abmm})$$

The L_0 part vanishes,

$$L^0_{abmm} = E_a\,\delta_{am}\,\delta_{bm} - E_m\,\delta_{mb}\,\delta_{am} = 0$$

This gives

$$K_{nnmm} = [L_1 e^{-it_1(1-D)L} \quad (1 - D)L_1]_{nnmm}$$

Note that it is formally of order λ^2

Next we prove the sum rule

$$\sum_m K_{nnmm} \equiv 0 \qquad \text{all n, all t}$$

This is because of

$$\sum_m K_{nnmm} = \sum_m\sum_c\sum_d \; [L_1 e^{-it_1(1-D)L} \quad (1 - D)]_{nncd} \cdot L_{cdmm}$$
$$= \sum_c\sum_d [L_1 e^{-it_1(1-D)L} \quad (1 - D)]_{nncd} \cdot \sum_m L_{cdmm}$$

But

$$\sum_m L_{cdmm} = \sum_m H_{cm}\;\delta_{dm} \cdot \sum_m \delta_{cm}\,H_{md} = H_{cd} - H_{cd} = 0$$

The sum rule can be used this way:

$$K_{nn,\,mm} = -\sum_{m \neq n} K_{nnmm}$$

and

$$\frac{d\rho_{nn}(t)}{dt} = -\sum_{m\neq n}\int_0^t dt_1\,K_{nn,\,mm}(t_1)\,[\rho_{mm}(t_1) - \rho_{nn}(t_1)]$$

This begins to get very close to a master equation. If we had symmetry of K, we could get even closer.

The symmetry

$$K_{nn,mm}(t) = K_{mm,nn}(t)$$

follows easily from the structure of K, as is often observed in such problems, in the absence of magnetic field. The argument is as follows:

Note first that

$$L_{abcd} = H_{ac}\,\delta_{bd} - \delta_{ac}\,H_{db},$$

$$L_{cdab} = H_{ca}\,\delta_{bd} - \delta_{ac}\,H_{bd},$$

so that

$$L_{cd,ab} = L^{*}_{ab,cd}$$

But in the absence of magnetic fields (and spin effects) we can always choose real unperturbed eigenfunctions. Therefore it is always possible to have a representation in which H_{mn} is real, and

$$L_{cdab} = L_{abcd}$$

Next, we note that the unit tetradic and the operator D are symmetric in the same way, e.g.,

$$D_{abcd} = D_{cdab}$$

But

$$K = DL_1\, e^{-it_1(1-D)L}\,(1 - D)L_1$$

and each operator appearing with K is symmetric to the exchange of the left and right pairs of subscripts. This K has the same symmetry,

$$K_{nnmm} = K_{mmnn}$$

(This result should also follow from more elegant time-reversal arguments. We do not pursue this here.)
Now our almost-master equation can be written in the form

$$\frac{d\rho_{nn}(t)}{dt} = -\int_0^t dt_1 \sum_m [K_{nn,mm}(t_1)\rho_{mm}(t-t_1) - K_{mm,nn}(t_1)\rho_{nn}(t-t_1)]$$

The only extra feature we have here is the convolution over time. This appears also in Van Hove's generalized master equation, as well as in those of Resibois and Montroll.

Note that we do not have to bring in the symmetry of K; there is no reason why we cannot keep it always as K_{nnmm}.

LECTURE III

The generalized master equation that was just derived reduces, in the limit of weak interaction, to the Pauli master equation. This can be seen in the following way.

The essential quantity is

$$K_{nnmm}(t) = [L_1 e^{-it(1-D)L} \quad (1 - D) L_1]_{nnmm}$$

Note that L_1 is of order λ. Let us take advantage of this explicitly by exposing all factors λ, or

$$L = L_0 + \lambda L_1$$

Then

$$K_{nnmm}(t) = \lambda^2 [L_1 e^{-it(1-D)(L_0+\lambda L_1)} \quad (1 - D) L_1]_{nnmm}$$

Now we introduce a new time scale,

$$\lambda^2 t = \tau$$

$$\rho_{nn}(t) = P(\tau)$$

Consequently,

$$\frac{\partial P_n(\tau)}{\partial \tau} = -\sum_m \int_0^{\tau/\lambda^2} dt_1 [L_1 e^{-it_1(1-D)(L_0+\lambda L_1)} \quad (1 - D) L_1]_{nnmm}$$
$$\times \{P_m(\tau - \lambda^2 t_1) - P_n(\tau - \lambda^2 t_1)\}$$

(We have not bothered here to invoke the symmetry of K.)

Now we just boldly take the limit $\lambda \to \infty$, for τ fixed. This gives

$$\frac{\partial P_n(\tau)}{\partial \tau} = \sum_m \int_0^\infty dt_1 \left[L_1 e^{-it_1(1-D)L_0} (1-D) L_1 \right]_{nnmm}$$

$$\times \{P_m(\tau) - P_n(\tau)\}$$

This limit is the same one used by Van Hove:

$$\left.\begin{matrix} \lambda \to 0 \\ t \to \infty \end{matrix}\right\} \qquad \lambda^2 t = \tau = \text{constant}$$

In the limit, we can return to the original notation,

$$\frac{\partial \rho_{nn}(t)}{\partial t} = + \sum_m W_{nm} \left[\rho_{mm}(t) - \rho_{nn}(t) \right]$$

where

$$W_{nm} = -\lambda^2 \int_0^\infty dt \left[L_1 e^{-it(1-D)L_0} (1-D) L_1 \right]_{nnmm}$$

Because of the limit, the integrand here contains

$$\exp - it(1-D)L_0$$

where L_0 is the <u>unperturbed</u> operator. This is a great convenience in continuing the calculation.

The simplification here is that D Lo vanishes:

$$(DL_0)_{abcd} = \delta_{ab} \left[H^0_{ac} \delta_{bd} - \delta_{ac} H^0_{db} \right]$$

$$= \delta_{ab} \left[E_a \delta_{ac} \delta_{bd} - E_b \delta_{bd} \delta_{ac} \right]$$

Therefore,

$$(1-D) L_0 = L_0$$

and

$$e^{-it(1-D)L_0} = e^{-itL_0}$$

But this gives the unperturbed time displacement operator, and we know how to handle it.

Recall from an earlier lecture that

$$(e^{-itL})_{mnm'n'} = (e^{-itH})_{mm'} (e^{itH})_{nn'}$$

For L = Lo, we get

$$(e^{-itL_0})_{mnm'n'} = e^{-itE_m} \delta_{mm'} e^{itE_n} \delta_{nn'}$$

$$= e^{-it(E_m-E_n)} \delta_{mm'} \delta_{nn'}$$

With this result we can work out all the operators in W_{nm}:

$$W_{mn} = -\lambda^2 \int_0^\infty dt \sum_{abcd} (L_1)_{nnab} (e^{-it(E_m-E_n)} \delta_{ac} \delta_{bd})$$

$$(L_1)_{cdmm}$$

Or,

$$W_{nm} = -\lambda^2 \int_0^\infty dt \sum_{ab} (V_{na} \delta_{nb} - \delta_{na} V_{bn}) e^{-it(E_a-E_b)}$$

$$x(V_{am} \delta_{bm} - \delta_{am} V_{mb})$$

$$= -\lambda^2 \int_0^\infty dt \sum_{ab} \Big\{ V_{na} V_{am} \delta_{nb} \delta_{bm} e^{-it(E_a-E_b)}$$

$$+ V_{bn} V_{mb} \delta_{na} \delta_{ma} e^{-it(E_a-E_a)}$$

$$- V_{na} V_{mb} \delta_{nb} \delta_{am} e^{-it(E_a-E_a)}$$

$$- V_{bn} V_{am} \delta_{na} \delta_{bm} e^{-it(E_a-E_a)} \Big\}$$

$$= -\lambda^2 \int_0^\infty dt \left\{ \sum_a V_{na} V_{an} \delta_{nm} e^{-it(E_a - E_m)} \right.$$

$$+ \sum_b V_{bn} V_{nb} \delta_{nm}\, e^{-it(E_m - E_b)}$$

$$- V_{nm} V_{mn}\, e^{-it(E_m - E_n)}$$

$$\left. - V_{mn} V_{nm}\, e^{-it(E_n - E_m)} \right\}$$

Because we need in general only $n \neq m$, the first two terms drop out. The second two terms combine to give

$$V_{nm} = +\lambda^2 |V_{nm}|^2 \int_0^\infty dt \left[e^{-it(E_m - E_n)} + e^{-it(E_n - E_m)} \right.$$

$$\left. = 2\pi\lambda^2 |V_{nm}|^2\, \delta(E_m - E_n) \right]$$

Which is precisely the Golden Rule formula.

Notice that a swindle has been perpetrated. Namely, we come out with a delta function in energy, without also introducing the concept of a continuous spectrum. For a finite system (finite volume, finite number of particles) the spectrum is discrete and the delta function as it appears here is meaningless.

There are two ways out. One is to take the limit of infinite system before taking the Van Hove limit

$$\lambda \to 0, \; t \to \infty, \; \lambda^2 t = \text{constant}$$

This appears to provide the necessary continuous spectrum.

The other way out, which is actually more informative, is to keep the finite system and to also keep t finite. Now we have to return to the non-Markoffian equation, but as we have already found K to the second order, we put it in.

$$\frac{d\rho_{nn}(t)}{dt} = \lambda^2 \int_0^t dt_1 \sum_m |V_{nm}|^2\, 2\cos(\omega_{mn} t_1)$$

$$\times \{ \rho_{mm}(t - t_1) - \rho_{nn}(t - t_1) \} + 0(\lambda^3)$$

Recall that the quantum number n really refers to the set of all quantum numbers needed to characterize the unperturbed state of a many-body system. The set n includes the unperturbed energy and also various occupation numbers, etc. Let's suppose that V_{nm} is a "smooth" function of these quantum numbers. Because

$$\frac{\partial \rho_{nn}}{\partial t}$$

is proportional to the "smooth" $|V_{nm}|^2$, we may also regard ρ_{nm} "smooth."

But there is one term in the master equation that is not smooth,

$$\cos(\omega_{mn} t_1)$$

If E_n is fixed, and we vary E_m, this will oscillate wildly if t_1 is large. Because of the oscillations, the sum over m will have a rather small value for large t_1. On the other hand, the sum will be large for small t_1.

What is large and small in this context?

Suppose that a typical value of ω_{mn}, for states for which $V_{nm} \neq 0$, is about $\bar{\omega}$.
Then large t_1 means

$$t_1 \gg \frac{1}{\bar{\omega}}$$

and small t_1 means

$$t_1 \ll \frac{1}{\bar{\omega}}$$

As an example, consider the scattering process in which two quasi-particles are annihilated and two others are created. Then $\hbar\bar{\omega}$ is of the order of a typical quasi-particle energy. For phonons in a crystal, $\bar{\omega}$ would be of the order of the Debye frequency. Anyhow, $\bar{\omega}$ is typically of an atomic order of magnitude, and not a macroscopic one. That means the important time scale for the "memory" t_1 is an atomic time scale.

But in a many-body system, $\rho(t)$ varies slowly with time; its own characteristic time scale is a macroscopic one—or at least one that is much longer than the previously mentioned atomic time scale. In fact, the time scale for $\rho(t)$ is inversely proportional to λ^2, and this has nothing to do with $\bar{\omega}$. Thus, for small λ, $\rho(t)$ will hardly change at all during the memory $1/\bar{\omega}$.

Because of this, we can safely approximate:

$$\rho_{nn}(t - t_1) \cong \rho_{nn}(t) \text{ for } t_1 \lesssim 0(1/\bar{\omega})$$

and

$$\frac{\partial \rho_{nn}(t)}{\partial t} \cong \lambda^2 \int_0^t dt_1 \sum_m |V_{nm}|^2 \cdot \cos \omega_{mn} t_1$$

$$x\{\rho_{mm}(t) - \rho_{nn}(t)\}$$

But now the t_1 integral can be done, giving

$$\frac{\partial \rho_{nn}(t)}{\partial t} \cong \lambda^2 \sum_m |V_{nm}|^2 \cdot 2 \frac{\sin \omega_{mn} t}{\omega_{mn}}$$

$$x\{\rho_{mm}(t) - \rho_{nn}(t)\}$$

For very large t the quality

$$\frac{\sin \omega_{mn} t}{\omega_{mn}}$$

looks very much like a delta function in ω; its actual width is of the order

$$\text{width} \sim 1/t.$$

We want to find many unperturbed states within this width, so that we can take limit of a large system with confidence.

In a many-body system the density of states is proportioned to the size of the system (either number of particles N or volume Ω.

The denisty of states is itself of order $1/\bar{\omega}$ because a typical ω_{nm} is of order $\bar{\omega}$. Therefore, the number of states in the interval $1/t$ is

$$\# \mathrm{int} = 0\left(N \cdot \frac{1}{\bar{\omega}} \cdot \frac{1}{t}\right)$$

and this should be much larger than 1 if we are to approximate $\frac{\sin \omega t}{\omega}$ by a delta function and still catch many states.

So, $\frac{N}{\bar{\omega} t} \gg 1$ or

$$t \ll N/\bar{\omega}$$

This limitation is not serious: by making N large, we can make the upper limit on t much larger than any time of experimental interest. The point is that, although t cannot actually become infinite as required in the Van Hove limit, nevertheless, the finite system, having N particles, behaves like the infinite system for a very long time. This is the justification for using the combination of delta function and continuous spectrum in the Pauli equation.

Finally, we note that the potential must have the property

$$\sum_{m} |V_{nm}|^2 \, x f_m = 0(N^0)$$

as $N \to \infty$. Here f_m is some arbitrary function of m. This requirement is a special case of what Van Hove calls the "diagonal singularity condition." Its consequence is that the transition rates W_{mn} are in fact independent of the size of the system-that is, the W_{mn} are intensive and not extensive properties. This expresses the macroscopic character of the processes described by the master equation.

In the derivation that I have just given, this diagonal singularity condition appears as a requirement that the results be physically sensible. In Van Hove's derivation, however, it is used to obtain the master equation in the first place.

Swenson has shown how Van Hove's derivation can be

carried out, like mine, without using the diagonal singularity condition until the very end, and then only for obtaining intensive transition rates.